Y0-BYV-053

# Science and Technology in the Development of Modern China

## *An Annotated Bibliography*

Genevieve C. Dean

SCIENCE POLICY RESEARCH UNIT
UNIVERSITY OF SUSSEX

Q180
C6
A12
D4
1974
O.C.

Mansell
1974

309192

*Research Aids of the East Asian Institute*
COLUMBIA UNIVERSITY

This bibliography is a part of the series *Research Aids of the East Asian Institute,* Columbia University, New York City. Preparation of the volume was undertaken between 1970 and 1972 with the aid of a grant from the Joint Committee on Contemporary China of the Social Science Research Council, New York, and the American Council of Learned Societies. A grant from the East Asian Institute has since enabled the revision and extension, with a view to making *Science and Technology in the Development of Modern China* accessible to a wider audience.

Copyright © 1974 Genevieve C. Dean
Mansell Information/Publishing Limited
3 Bloomsbury Place, London WC1A 2QA, England

International Standard Book Number 0 7201 0376 2
Library of Congress Card Number 74-76296

Printed in Great Britain by Kingprint Ltd., Richmond, Surrey

# Contents

Science, Technology and Development

Introduction to the Bibliography

Bibliography

I *Technology and Economic Growth* 1

II *Technology Policy* 9
- A Choice of Techniques 17
- B Transfer of Technology 22
- C Innovation and the Direction of Technological Change 25

III *Science Policy* 37
- A Policy for Science 42
- B Policy toward Science 66

IV *Scientific Activities* 75
- A Research and Development (R & D) 80
  - 1 The Natural Sciences 80
  - 2 Industrial R & D 89
  - 3 Agricultural R & D 89
  - 4 Medical R & D 90
  - 5 Military, Space Exploration and Travel, and Nuclear Energy R & D 91
- B General Purpose Data Collection 100
- C Scientific Communication 100
- D Higher Education and Technical Training 106
- E Scientific and Technical Education at the Primary and Secondary Levels; Extension of Scientific and Technical Information; Popularization of Science 118
- F Testing and Standardization 119

V *Technology in China* 121
A Agricultural Technology 125
B Industrial Technology 131
1 Electric Power Industry 131
2 Electronics Industry 132
3 Chemical Industry 136
4 Petroleum Industry 138
5 Textile Industry 145
6 Machinery Industry 145
7 Transport Equipment Industry: Motor Vehicles 149
8 Basic Metals Industries 150
9 Mining Industry 153
C Engineering 158
D Transportation and Communications 160
E Natural Resources 162
F Trade 165
G Technology and Employment 168
H Medicine and Public Health 171
I Population Control 179
J Environmental Control 182
K Technical Assistance from China 184

Appendices
I Modern Science and Technology in China before 1949 188
II Traditional Chinese Science and Technology 201

Author Indices
I Primary Material 226
II Secondary Material 231
III Tertiary Material 262

# Science, Technology and Development

The framework of this bibliography is based on two premises. The first holds that an important source of economic growth is the substitution of more efficient for less efficient technologies of production—in other words, a change in the combination of capital, land and labour which enables a larger output to be produced with the same quantity of investment. Because the proportion of economic growth which can be attributed to technological change is difficult to measure it is often treated by economists as a residual, or as that part of the increased product which cannot be accounted for by larger investments of capital and labour. Economic growth may also be the consequence of technological changes which make it possible to exploit hitherto unutilized resources, or to produce new goods or new qualities of goods which open up new markets.

The second premise underlying the study of science, technology and development holds that modern science and the methods of scientific research and experimental development (R & D) are universally valid sources of new technologies and technological improvements. 'Science policy' can refer to 'research policy'—that is, to the allocation of financial resources and manpower; the organization and managements of institutions for research; and the scientific activities which support R & D such as collection of data, communication of scientific and technical information, education and training of scientific and technical manpower, etc. 'Science policy' can also refer to the application of scientific research to economic and social problems and to the creation of new scientific theories for their solution. The application of the natural sciences (with which we are concerned here, to the exclusion of the social sciences) to social needs occurs through research, leading to the invention of new technologies: 'social technologies' such as new medical techniques, military technologies, and the development of these technological inventions into practicable machines and processes. Science policy studies were originally concerned with policies for research and education, thus tacitly equating the growth of R & D capabilities with the initiation of technological advance in the desired direction. It has been recognized more recently that the growth of scientific research has not in itself guaranteed technological, and hence economic and social, development; therefore contemporary studies of science policy attach greater importance the actual application of scientific knowledge to economic and social problems. This bias in science policy carries significant implications for research policy and the direction of research, and for the relative importance of research and of the other scientific and engineering activities which make up the experimental development stages of the innovation process.

Thus the study of 'science, technology and development' is concerned with the role of technological advance in economic and social development, and with the sources of technological innovation. While many of the concepts used have been borrowed from economics, the study of technological change is interdisciplinary and includes analysis of the effects of political and social structures, of economic factors, and of technical-engineering constraints on the 'demand' for and 'supply' of new technologies.

The specific nature of the problems and the precise goals of development vary from country to country. Nevertheless, it is possible to define development, for present purposes, in terms broad enough to encompass both the objectives of development policies in the Peoples' Republic of China and in countries with widely differing social, economic and political structures. We are less interested here in specific economic and social objectives, than in the means by which these countries commonly aim to attain their development goals. 'Development' is used here to refer to economic growth resulting from the use of more efficient technologies of production rather than merely from the investment of more capital, land or labour. Sustained economic growth and development implies that progressively more efficient technologies are continually replacing less efficient ones; and this in turn presupposes that there is routine access to a source of new technologies and that there exist economic and/or social structures which induce technological innovation.

'Development' also implies that the material wealth which is created by the growth of output is ultimately used for improvements in material standards of well-being, distributed on an increasingly egalitarian basis among the population of a developing country. It is the addition of such social objectives which distinguishes 'development' from mere 'growth'. A further qualification to our definition of 'development' is that the technological change which allows the production of greater output, must occur in response to demands arising within the economy of the developing country itself: this excludes technological changes occurring in a colonial economy as a result of economic expansion and market forces operating in other countries. Furthermore, it allows consideration of the fact that the 'demand' for technological change does not necessarily contribute to the goal of wider income distribution even when that demand arises within the country in question: if a particular economic, social or political group controls access to new technologies through its domination of local R & D institutions, and its monopoly of foreign exchange and of licenses to use foreign technologies, then, in the absence of other social or economic mechanisms to redistribute the increased product, innovation may simply increase the wealth of that group.

The 'development' objective shared by China and other developing countries is the grounds on which a comparative approach was chosen for this bibliographic survey. Technological advance as the basis of economic growth, and scientific research as the basis of technological advance, are the instruments used by China and by many countries for attaining the objective of 'development'.

The study of 'science policy' is a comparatively new field, particularly when undertaken in the context of development. One purpose of this bibliography is to introduce China specialists to the broad concepts of 'science, technology and development' via literature with which they may already be familiar, but which is reviewed here from the perspective of science policy. In the Chinese literature itself—notably in policy documents and the statements of individual political leaders—technological advance is clearly regarded as the key to industrial, agricultural and military modernization, and modern science is intended to be the basis of technological modernization and fundamental social and cultural change. The root of many policy conflicts in China lies in differences of opinion about the direction of technological change—not about the ultimate objective of such change, but in differences about the starting point from which change toward technological modernization can begin. These differences lie in the different time horizons of planners, and in their respective priorities for short- and medium-term development objectives, whether technological, economic or social. Must technological modernization of the Chinese economy await construction of a modern capital goods industry in China, capable of supplying new machines and equipment to agriculture, light industry and heavy industry? Or should some investment be made in raising the current levels of productivity in agriculture and industry short of completely replacing the means of production they use? Such differences of opinion also seem to underlie the apparent conflicts over policies for science. The growth of modern science is expressed either in the growth of the capacities of conventional scientific institutions, followed by the gradual extension of the fruits of scientific research through society; or in the widespread diffusion of knowledge of the basic concepts of science throughout the population, so that social and cultural change occur in step with the gradual development of scientific capabilities. Analysis of the economic and technical constraints in various strategies of scientific and technological development, can therefore enhance our understanding of Chinese behaviour in a much wider area than merely in the area of science and technology policy.

Despite the existence of a large body of relevant literature there are few studies of Chinese science and technology, or of economic development in China, with this particular orientation. Students of science policy have been deterred from working on Chinese science, technology and development by the need for specialized research techniques to cope with the type of data available. This bibliography assembles the existing literature on these central aspects of the Chinese experience for such scholars.

Chinese policies for technological modernization may contribute a valuable perspective to science policy and development studies, particularly in the choice of techniques and its ramifications for the direction of technological change and for the sources of appropriate new technologies. Modern science in China, as reflected in the organization and support of scientific institutions and activities, appears to be fairly conventional. What may be unique in the Chinese experience however, is the way 'science' has been applied to technological advance in the context of 'walking on two legs'—that is, the choice of different kinds of technology for different economic sectors. Are, in fact, the conventional institutional structures, patterns of organization and methods of procedure in modern science universal and fixed; is science in its conventional forms applicable to the development of small-scale, labour-intensive technologies? Or have unique scientific institutions and methodologies been developed in China on the fringes of a conventional science system, in order to direct part of the scientific effort toward the technological requirements of that sector? If so, what are these new institutions, and what is the relationship between them and the scientific establishment? What is the relationship between the two 'legs' of technology: do they compete for scientific, technical and engineering resources, or are there specific provisions in Chinese science policy which make simultaneous technological advance in both sectors complementary? Finally, what are the instruments through which science and technology policies are implemented in China—are they explicitly for scientific development and technological advance, or are they implicit in economic, political, social and military policies? Which of these policies and instruments of policy are applicable in other developing countries, and what can other countries—developed and developing—learn from the Chinese experience? It is hoped that from the discussion between specialists in the study of science policy, and students of contemporary China, there will emerge a general recognition of the potential utility of the science policy framework for analysing China's development, and some understanding of the ways in which knowledge of the Chinese experience can contribute to the study of science, technology and development.

# Introduction to the Bibliography

This bibliography was originally compiled for a Study Group on Science and Technology in China's Development, held at the University of Sussex in January 1972.[1] Its purpose was to identify and catalogue Western literature pertaining to modern science and technology in China: in effect, to describe the 'state of the art' in preparation for the Study Group's discussion of future research priorities in the field of Chinese science policy studies. It originally included about 800 entries, listing books, articles and unpublished papers and dissertations, with annotations to show their relevance to the study of the role of scientific and technological advance in the economic and social development of China. The arrangement of entries in the bibliography has subsequently been revised in light of the Study Group's conclusions, and coverage extended to include items published during 1972.

The organization of the entries is an adaptation of the classification scheme prepared for a bibliography on science, technology and development compiled at the Science Policy Research Unit, University of Sussex, in 1969.[2] The original classification had categories for conventionally defined scientific activities; fundamental issues in policy-making for science and

1 The Study Group on Science and Technology in China's Development was convened from 10 to 14 January 1972 by the Science Policy Research Unit, and was sponsored by the Joint Committee on Contemporary China of the Social Science Research Council and the American Council of Learned Societies and by the International Development Research Centre (Canada). The discussion at the Study Group are described by Genevieve Dean in *New Scientist* (see entry no. 5) and *The China Quarterly* (entry no. 6), and by Susan Rifkin in *Science Studies* (entry no. 15) and *Items of the Social Science Research Council* (entry no. 16).
The papers prepared as background for the Study Group have not been published as a collected study, but some were subsequently revised for individual publication. These include the articles by Shigeru Ishikawa (entry no. 29), by Jon Sigurdson (entry no. 107), by Genevieve Dean and Manfredo Macioti ('Scientific Institutions in Contemporary China', *Minerva,* (in press), and by Susan Rifkin and Raphael Kaplinsky ('Health Strategy and Development Planning: Lessons from the People's Republic of China', *Journal of Development Studies,* 9, no. 2 (1973), 213-232.

2 G. C. Dean, et al. *Bibliographic Study of Science, Technology and Development,* unpublished mimeographed and punched cards, prepared with the aid of a grant from the Canadian International Development Agency.

technology; the social and economic sectors in which technology is used and in which technological change is necessary for development; and various codes for indicating the geographical reference, the type of literature and the academic discipline. With some modification to suit both the particularities of the Chinese case and the format of this bibliography, the scheme has been retained in order to facilitate comparison with the general bibliography on science, technology and development.

Because none of the items listed originates from Chinese sources, this bibliography must be defined as a survey of 'secondary' materials. Some of the references, however, are 'primary', in that they record the direct observations of the authors; others are 'secondary', in the sense of being analyses and critiques of primary information; still others are 'tertiary' in nature, being reviews and commentaries on the secondary literature, bibliographies and 'aids to research' in the field. An alphabetical author index, arranged according to these three categories, has been supplied to assist the reader in selecting items appropriate to his requirements.

This listing of items on science and technology in China expands and updates the 1966 publication by the Research Policy Library (Lund) and Research Survey and Planning Organization (New Delhi), *Research Potential and Science Policy of the People's Republic of China; A Bibliography, (see* entry no. 186). Other more specific bibliographies which were consulted are cited in the text. The annual bibliographies of the *Journal of Asian Studies* and standard periodicals in the field of China studies were also searched. Other articles were found in serial publications in the field of science policy studies and in scientific, technical and industrial journals. Most of the literature was searched systematically back to 1960, but it should be noted that the *Far Eastern Economic Review* was searched only back to 1965, and only major articles have been cited. The multi-disciplinary character of this field of study is reflected in the sources on which this bibliography draws.

Books and articles from British, American, Japanese and European publications are cited, as well as some items translated from the Russian. A number of unpublished papers which have been located and found to be of particular relevance are also included. The reader is referred to the following bibliographies for unpublished doctoral dissertations and works-in-progress:

Leonard H. D. Gordon and Frank J. Shulman, *Doctoral Dissertations on China: A Bibliography of Studies in*

*Western Languages, 1945–1970,* University of Washington Press, Seattle, 1972.

Curtis W. Stucki, *American Doctoral Dissertations on Asia, 1933–June 1966,* Data Paper 71, Southeast Asia Program, Department of Asian Studies, Cornell University, October 1968.

*Tung-li Yuan, A Guide to Doctoral Dissertations by Chinese Students in America, 1905–1960,* Sino-American Cultural Society, Inc., Washington, 1961.

and to issues of:

*Dissertation Abstracts,* University Microfilms, Inc., Ann Arbor, Michigan.

*Modern China Studies,* International Bulletin, Published bi-annually by *The China Quarterly* since August 1970.

*Bulletin de Liaison pour les Études Chinoises en Europe,* published by École Pratique des Hautes Études, Paris.

*Asian Studies Professional Review,* published by the Association of Asian Studies.

This bibliography is believed to be a comprehensive listing of 'secondary' items on science, technology and development in the People's Republic of China. The history of science and technology in China does not, of course, begin with 1949 and the legacy of both pre-existing modern scientific institutions and capabilities and of traditional sciences and technologies has been an important factor in development policies since 1949. During the preparation of this bibliography, many items were found which describe and analyse this legacy. These have been listed in two appendices—one on traditional Chinese science and technology and one on the introduction of modern science and technology to China and its history prior to the People's Republic. Because these are probably representative, rather than exhaustive listings of the literature, these aspects of Chinese science and technology have been treated as appendices to the main body of the bibliography. For further references the reader is advised to consult John Lust, *Index Sinicus* (entry no. 798) and the annual issues of *Isis: Critical Bibliography of the History of Science and its Cultural Influences* (entry no. 862).

This bibliography was undertaken in order to determine the 'state of the art' in studies of science and technology in China's development. A much greater quantity of literature was discovered than had been anticipated, representing a considerable body of information and knowledge about science and technology policies in China, but the general conclusion drawn from this survey is that the role of science and technology in Chinese development has not been fully assessed in the existing literature. This is partly due to the scarcity of concrete data on the Chinese experience. But, as some of the entries in this bibliography demonstrate, there is sufficient information available to begin formulating hypotheses about Chinese science and technology policies, even though opportunities to test these hypotheses empirically are still limited. With relatively few exceptions, this literature only begins to exploit the potential value of studies of Chinese science and technology policy. The Study Group for which this bibliography was prepared concluded that:

> In three respects, the Chinese case can be especially instructive. First, by analyzing scientific and technological development in the terms the Chinese themselves use, contrasting the Chinese view of their own development with their views on development elsewhere and with our own understanding of development, we may acquire a much richer understanding of this process. Secondly, by using the Chinese case to construct the kind of comprehensive framework in which economic, technical and scientific decisions, and the consequences of those decisions, can be linked to the institutional and organizational setting, we may improve our understanding of our own system. Finally, the particular study of Chinese labour-intensive technological change may prove of exceptional interest for many developing countries experiencing high rates of unemployment.

It is hoped that by drawing the attention of a wider audience to the existing literature, this bibliography will contribute to these aims.

# SECTION I

## *Technology and Economic Growth*

The items listed in this section are general studies of economic development and strategies of industrialization in China since 1949. These articles reflect the fact that technological innovation has been recognized by economists for some time as an important source of growth in the industrialized market economies. Increasingly in these societies the relationships between science and technology and between research and innovation, have influenced innovativeness in their economies. Economic 'underdevelopment' was at first understood to result from the lack of efficient modern production technologies, and 'development' to follow from the transfer of modern technology from the industrialized to the 'underdeveloped' countries. But, by the 1960s however, increasing numbers of economists held that economic development required an indigenous R & D capability in order to sustain a process of technological innovation. In an underdeveloped economy however, the critical problem may often be the accumulation of an economic surplus to invest in new technology and in scientific resources—a process of capital accumulation which, without outside assistance, depends on precisely the economic growth it is intended to generate. The issues to be examined in this context thus include the method of initial capital formation for industrialization, the acquisition of industrial technologies and improved agricultural technologies, and the stimulation of a continuous process of technological innovation as the basis of ongoing economic growth.

The items in this section are concerned with development chiefly in the sense of economic growth: they exclude the social objectives which have been used to qualify the definition of the term in this bibliography. The articles study economic growth in terms of increasing the quantity of material output which is obtained from the production process. Most refer to technological advance in China as a consequence rather than as a means of economic growth, and consider the process of technological innovation and the sources of new technology, only in passing, if at all.

1 ADLER, Solomon
*The Chinese Economy*. London, Routledge & Kegan Paul, 1957.
General economic survey. Technical training and manpower, p. 66, 77, 97, 201. Strategy for technological modernization: 'the gap between modern technological standards and best Chinese practice is being narrowed'; 'the considerable opportunities for advance in labour-intensive, smaller-scale and even petty production are not being neglected', p. 93. Agricultural technology, p. 117-27. Soviet technical assistance, p. 57, 88.

2 ASAHI NEWSPAPERS (ed.)
Science and technology in modern China. *The Asahi Asia Review*, 3(1), 10-61, 1972.
Reiitsu Kojima--Mao Tse-tung's thoughts on science and technology. Shuzo Yamato--Present and future of the machine tool industry. Saburo Tamura--Chemicals and mechanization processed by the mass movement of workers and peasants. Masumi Sato--Mining industry, including the history and future of the iron and steel industry. Shigeru Tsutsumi--From natural to man-made fibres. Nobuhisa Akabane--Chemical industry: Technological self-reliance preceded the development stage. Hiroshi Ito--Development of the marine industry and the canal system. Kyoichi Harashina--Growth centered around tractor building. Akio Akagi--Large-scale electronic computers for general use. Etzuso Onoe--Energy policy: The pattern of development in the 1970s. Jugo Wakamatsu--Powerful new weapons are mass-produced to China's own designs.

3 BUCK, John Lossing, DAWSON, Owen L., and WU, Yuan-li
*Food and agriculture in Communist China*. New York, Praeger for the Hoover Institution, 1966.
From 1949 to 1960, Peking followed an agricultural policy which was aimed at self-sufficiency in food; reversal of this policy in 1960 was a matter of political necessity which grew out of a severe food shortage . . . however, self-sufficiency in food supply remains one of the major objectives of the Communist regime. John Lossing Buck--Food grain production in mainland China before and during the Communist regime. Yuan-li Wu--The economics of mainland China's agriculture: Some aspects of measurement, interpretation, and evaluation (long-term trend of China's grain production, based on an 'adjusted model' of grain production, predicted upon gradual improvements on the existing technological base). Owen L. Dawson--Fertilizer supply and food requirements; Irrigation developments under the Communist regime (the core of the technical problems).

4 CHEN, Nai-ruenn and GALENSON, Walter
*The Chinese economy under Communism*. Chicago, Aldine, 1969.
General economic survey by chapters: 1--The economic heritage. 2--Alternative paths to economic development. 3--Development of the industrial sector (the Soviet assistance program, p. 51-5; the choice of techniques in manufacturing, p. 69-72). 4--Agriculture. 5--Population and employment. 6--The control and allocation of resources. 7--Conditions of life and labor. 8--Foreign economic relations; the commodity composition of foreign trade, p. 203-6; the Chinese foreign aid program, p. 212-4. 9--Prospects for the Chinese economy.

5 DEAN, Genevieve
China's technological development. *New Scientist*, 54(796), 371-3, 1972.
The applicability of Chinese policies for technological development to other developing countries. Based on the Study Group of Science and Technology in China's Development, University of Sussex, January 1972.

6 DEAN, Genevieve
Science, technology and development: China as a 'case study'. *The China Quarterly*, No. 51, 520-34, 1972.
Summary and implications of discussions at Sussex Study Group on Science and Technology in China's Development, January 1972. Choice of technology; rate and direction of technological change; sources of technological change; science policy. Priorities for research in the field.

7 DONNOTHORNE, Audrey
*China's economic system*. London, George Allen and Unwin Ltd., 1967.
General economic survey. Includes references to technical manpower, technical innovations, technical standards, organization of industrial and academic research, science and scientific personnel, organization of science.

8 DUNCAN, James S.
Red China's economic development since 1949. *Contemporary China*, Ruth Adams (ed.). New York, Vintage Books, p. 139-49, 1966. Also *Bulletin of the Atomic Scientists*, 22(6), 84-7, 1966.
Progress in all fields 1949-59; catastrophic breakdown in China's economy, 1960. Chinese scientific development in metallurgy, radar and instrumentation is noteworthy; progress in nuclear science. Trade with West has brought modern industrial and scientific equipment. Role of students, scientists and technicians in economic development.

9 ECKSTEIN, Alexander, GALENSON, Walter and LIU, Ta-chang (eds.)
*Economic trends in Communist China*. Chicago, Aldine, 1968.
General economic survey. 4--John S. Aird--Population growth. 5--Chi-ming Hou--Manpower, employment and unemployment. 7--Anthony M. Tang--Policy and performance in agriculture. 9--Kang Chao--Policies and performance in industry (concerning industrial technology). 11--Robert Michael Field--Labor productivity in industry.

10 ISHIKAWA, Shigeru
The Chinese economy: A general framework for long-term projection. *The China Mainland Review*, 2(2), 71-90, 1966.
Discussion of the features of a development model for long-term projection of the Chinese economy. 'A basic hypothesis in our study is that changes in the economic development of China can be studied within a comprehensive framework of analysis applicable to all contemporary underdeveloped countries.' Five initial conditions and strategies in China: the supply and demand of labour force; agriculture; transplantation of technologies;

restrictive export markets; institutional setting. Features of the projection model.

11 ISHIKAWA, Shigeru
*Long-term prospects for the Chinese economy.* Tokyo, Asian Economics Research Institute (Institute of Developing Economies), No. 76 (Pt.I), 1964; No. 102 (Pt.II), 1966; No. 119 (Pt.III), 1967; Nos. 171-2 (Pt.IV), 1970.
No. 76: IV--Masumi Sato--China's technological level and outline of development; VI--Yoshio Akeno--The Chinese iron and steel industry; VII--Etzuso Onoye--The Chinese electric power industry. No. 102: VI--Masumi Sato--Chinese medium and small-scale industry seen from the technological angle; VII--Nobuhisa Akabane and Reiitsu Kojima--China's chemical industry; VIII--Reiitsu Kojima--The Chinese machine tool industry; IX--Etzuso Onoye--The Chinese power industry (supplement by Nobuhisa Akabane); X--Yoshio Akeno--The Chinese transportation industry; Discussion meeting--technological special features of Chinese medium and small-scale industry. No. 119: III--Nobuhisa Akabane--Consideration of methodology in the analysis of technological levels; IV--Fumio Kobayashi--Organization of scientific and technical research; V--Masumi Sato--China's technological development in the last ten years; VII--Reiitsu Kojima--The agricultural machinery and implement industry. No. 172: VI--Shigeru Ishikawa--Long-term outlook for Chinese agriculture; VII--Reiitsu Kojima--The iron and steel industry; VIII--Masumi Sato--Course of development of the electronics industry; IX--Reiitsu Kojima--Textile industry.

12 LI CHO-MING
China's industrial development, 1958-63. *The China Quarterly*, No. 17, 3-38, 1964.
General economic survey. China is growing more independent in technological matters. Modern industry vs. handicrafts. Soviet aid.

13 MOORSTEEN, Richard
Economic prospects for Communist China. *World Politics*, 11(2), 192-220, 1959.
Economic development of China: 'the outlook...makes sustained, rapid growth seem entirely possible'. The Soviet pattern of development; Soviet assistance and trade with the Soviet Union. Section on small-scale. labor-intensive industry, p. 203-6.

14 ORCHARD, J.E.
Industrialization in Japan, China mainland, and India, some world implications. *Annals of the Association of American Geographers*, 50(3), 193-215, 1960.
General survey of economic development in Japan, India and China. Problems of industrialization: acquisition of technical and administrative skills; limitations on capital formation (partially offset in China through mass organization of surplus labour for public works, emphasis on small-scale enterprises and labour-intensive methods); supplies of energy resources and raw materials. Evaluation of the progress of industrialization, China mainland: 'Walking on two legs' explains in part China's 'truly phenomenal' rate of economic growth.

15 RIFKIN, Susan B.
On 'contradictions' among academics (a commentary on a workshop). *Science Studies*, 2(4), 395-9, 1972.
'Contradictions' in discussions at the Sussex workshop on science and technology in China's development (Jan 1972) arising from different backgrounds of the participants.

16 RIFKIN, Susan B.
*Science and technology in China's development: report on a workshop held with the aid of the Joint Committee on Contemporary China.* Social Science Research Council, 26(1), 4-6, 1972.
Areas of future research on Chinese science policy, science and technology, as identified at the Study Group on Science and Technology in China's Development, University of Sussex, 10-14 January 1972.

17 ULLERICH, Curtis
Size and composition of the Chinese G.N.P. *Journal of Contemporary Asia*, 2(2), 163-80, 1972
'Thus, the official estimates of China's GNP advanced by the official or unofficial advisers of the major Western governments, whose posture is characterised by a negative attitude in principle, agree . . . (on) a gradual and inescapable shrinking of the per capita GNP . . . . 'Let us look at the parameters one by one and see whether these allegations are likely to correspond to the facts.' Agriculture: modern technological inputs; industry: 'China's industrial sector is composed of two structurally very different components . . . one . . . composed of large and larger-than average units of modern and very modern technological standards . . . . (The other) sector is composed of smaller-than-average and often very small units of intermediary and low technological standards, although the trend to rapid improvements in technology and output quality is very strong.'

18 U.S. CONGRESS: JOINT ECONOMIC COMMITTEE
*An economic profile of mainland China.* New York, Praeger, 1968.
Edwin F. Jones--The emerging pattern of China's economic revolution, p. 77-96. K.P. Wang--The mineral resource base of Communist China, p. 167-95. Marion R. Larsen--China's agriculture under Communism, p. 197-267. John S. Aird--Population growth and distribution in mainland China. Leo A. Orleans--Communist China's education: Policies, problems, and prospects, p. 499-518. Chu-yuan Cheng--Scientific and engineering manpower in Communist China, p. 519-47. Leo A. Orleans--Research and development in Communist China: Mood, management, and measurement, p. 549-78. Milton Kovner--Communist China's foreign aid to less developed countries, p. 609-20.

19 U.S. CONGRESS: JOINT ECONOMIC COMMITTEE
*People's Republic of China: An economic assessment.* Washington, U.S. Govt. Printing Office, 1972.
Philip D. Reichers--The electronics industry of China. Alva Lewis Erisman--China: Agricultural development, 1949-71. Philip W. Vetterling and James J. Wagy--China: The transportation sector, 1950-71. Leo A. Orleans--China's science and technology: Continuity

and innovation. John S. Aird--Population policy and demographic prospects in the People's Republic of China. Leo Tansky--Chinese foreign aid.

20 WHEELWRIGHT, E.L. and MCFARLANE, Bruce
*The Chinese road to Socialism.* New York and London, Monthly Review Press, 1970.
The main issues we were trying to understand were: 'First, the broad political, social and cultural factors which are so important in the Chinese economy. Second, how these operate in detail . . . . Our approach was conditioned by the conviction that it is not sensible to look at China through the prism of Western textbook economics . . . . Our approach was to examine the institutions which the Chinese state has set up to solve its economic problems and to distribute the power of decision-making. Moreover, we became aware . . . that the performance of the Chinese economy cannot be judged by purely economic standards . . . because non-economic aims are being "fed" into the planning system.' The authors visited China in 1966 and 1968. Chapter 9--Technological policy: The milieu of technology, an economy distinguished by social control and by the discipline of the unified state plan, to which is attached the 'motor' of moral incentives; the practice, descriptions of case studies; the theory, 'why have the Chinese been content to pursue a policy in which self-reliance, medium-scale technology . . . are pursued at the cost of lower productivity?'; technological achievements, case examples; manpower and science, science and economic planning; technology and the revolution in education; problems, the main problems appear to lie in technological policy.

21 WILSON, Dick
China's economic prospects. *Contemporary China,* Ruth Adams (ed.). New York, Vintage Books, P. 179-93, 1966.
Three handicaps to Chinese economic growth: technology (it is questionable whether China can achieve fast industrialization without know-how from a highly industrialized country), politics (political idealism may be particularly damaging in the field of science), population explosion.

22 WU, Yuan-li
Chinese industrialization at the crossroads. *Current History,* 41, 151-6, 1961.
General discussion of the status of the Chinese economy in 1960. Main points of current economic policy include emphasis on technological innovation, cost reduction, increase in labor productivity, quality improvement and the introduction of new products. Technical constraints dictate labor intensive methods, small industrial units.

23 WU, Yuan-li
The economy after twenty years. *Communist China, 1949-1969: A twenty-year appraisal,* Frank N. Trager and William Henderson (eds.) Chap. 6. New York, New York University Press, 1970.
P. 135-6: Success in selected industries versus failure in achieving technological independence. 'Breakthroughs on such a selective basis were at the expense of other industries and, indirectly if

not directly, of personal consumption as well . . . although the entire Chinese experiment in economic development has been financed by domestic savings alone . . . . Peking's success in industrial development during the early years of this period was to a considerable extent due to Soviet aid in equipment and technology.' P. 143-4: Conditions for economic growth. Favorable conditions include the resource base (new discoveries of natural resources; increased capital stock) and the level of Chinese technology; developments in rail and road transportation; 'tremendous' expansion of the research and development industry. Estimates of expenditures on science. 'Because of the long time-lag between research and application, China may be rapidly reaching a point in time when the positive results of indigenous research could begin to be felt.'

24 WU, Yuan-li
*The economy of Communist China; An introduction.* New York, Praeger, 1965.
Science and technology in the economic setting. Economic planning organs: 'the Scientific and Technological Commission is in charge of the collection of technical information from abroad and the dissemination of new knowledge', 'the Scientific and Technological Commission, the Bureau of Standards, and the Academy of Sciences form the backbone of China's R&D sector', economic development and technological improvements: 'the national economic plan must provide for the introduction from abroad of new knowledge, new methods of production, and new equipment, as well as for the encouragement of R&D at home.' Components of the national economic plan include a technological plan. Import of industrial equipment.

25 WU, Yuan-li
Industrialisation under Chinese Communism. *Current History*, 39(232), 343-9, 366, 1960.
The most important single factor in determining the future of the Communist economic program during the next few years will be the degree of success of the technological revolution.

26 WU, Yuan-li
Die zukünftige wirtschaftliche entwicklung in China. *Umschau in Wissenschaft und Technik*, No. 9, 1970.
Four positive factors in the future economic development in China are: the discovery of mineral resources as a result of geological surveys; improvement of transport facilities; the improvement of agricultural output; and the improvement of Chinese science and technology, which can result in dramatic rise in productivity through use of revolutionary methods of production. In agriculture: chemical fertilizers, insecticides, selection of high yielding plants; in the industrial sector, great efforts were made after the withdrawal of Soviet specialists to broaden technical imports and to hasten technical self-sufficiency through import of plant for the key industries and the construction of engineering institutes to copy this plant. Trade with Japan, West Germany, Italy, Great Britain and France. Growth of technical manpower.

# SECTION II

## *Technology Policy*

### A *Choice of Techniques*

The choice of techniques is perhaps the most critical issue which faces science and technology policy-makers, economic planners or entrepreneurs in a developing country. A rational choice among alternative technologies must take into consideration endowments of natural resources, the relative supplies of capital and labour, market conditions which constrain the nature of the product, as well as non-economic, social objectives such as employment, industrial health and safety, and pollution. Technical choice also involves the question of sources of innovation: where do techniques with the specified characteristics come from? Do they already exist, or must they be invented? How can the invention of appropriate techniques be stimulated? The choice of techniques has long-range implications for development, if the planners' objective is to raise prevailing levels of productivity in a systematic fashion. If the initial investment is in relatively inefficient techniques with low levels of productivity, then it is necessary to determine what economic, technical, and social conditions will compensate for the foregone product or economic surplus, and how innovation can be induced in order to raise the productivity of these techniques.

In economic theory, 'choice of techniques' refers to identification of the technique which makes the most efficient use of the factors of production: the technique which uses less labour or less capital than any other to produce a given output. The theoretical range of equally efficient alternative techniques, the 'production function', includes techniques which use different ratios of capital to labour. The individual entrepreneur or economic planner who makes investment decisions, must consider the relative supplies and hence the relative prices of capital and labour in the economy, and choose the technique which uses these factors of production accordingly. The addition of social objectives such as full employment to these purely economic considerations imposes further constraints on the choice of techniques.

Economic theory implies that the planner has access to a number of techniques of production which differ in the ratios of capital and labour which they use; in practice, however, the range of alternative techniques of equivalent efficiency may be considerably narrower than is implied by the theory. One objective in the science policy of a developing country may be to expand the range of practicable technological alternatives so as to increase the opportunity of choosing the techniques suited to the relative supplies of capital and labour in its economy. The purpose of studying policies for science *and* technology is to identify the obstacles in the way of realizing potential techniques which are feasible, given current levels of engineering knowledge, but which have not been developed into machinery and equipment, processes of production, or economic supplies of raw materials. If the role of economic analysis is to describe appropriate technologies, the role of science and of engineering is to invent techniques answering these descriptions, and the task of the student of science policy is to investigate why such techniques very often are not invented by scientists and engineers in the developing country.

Although 'invention' may theoretically occur in any economic or social environment, in practice 'innovation'—the use of a new invention—is observed to take place in response to specific needs or to perceptions of potential advantage or gain. The 'demand' for innovation and the technical change itself can occur in two separate economies, as for example in the introduction of industrial technologies from an advanced economy into a colonial or semi-colonial economy. When this happens, the economic surplus which is produced as a result of the innovation is not necessarily directed toward the social objectives specified by 'development', as the term is used here, and may not even be re-invested within the colonial economy. Furthermore, the effective demand for innovation may sustain a bias toward technologies which are relatively inefficient or are inappropriate for achieving economic and social development, if the domestic and international markets for the developing country's goods predispose its entrepreneurs to the use of such technologies. Thus, the choice of techniques is not only an economic and engineering problem, but also depends on the structures which control the market, and the economic, social and political forces to which these structures respond. Therefore, in order to give precedence to development objectives over the profit-maximizing goals of the private entrepreneur, the choice of appropriate techniques is now often seen as the outcome of a planning process rather than a consequence of the free operation of market forces.

The first four items listed in this section attempt to apply and develop the economic concept of choice of techniques as a tool in analysing Chinese economic policies, particularly in the area of industrial development. The other entries in this category are concerned with different economic strategies which have prevailed in different phases of Chinese policy, and the conflict between proponents of the different strategies. Each of these strategies bears certain implications for the choice of techniques and the direction of technological change in China.

Having made a choice of techniques—that is, determined the relative amounts of labour and capital to be invested in a production process—the planner or the entrepreneur must then find an efficient technology which uses the factors of production in the prescribed ratio. The way in which surplus product is translated into new and more efficient machinery, tools and equipment, and into the discovery and use of new raw materials and sources of energy which can be supplied more cheaply to manufacturers, refers to the problem of the sources of technological change. Technological change is represented by qualitative changes in the factors of production, particularly by changes in capital goods and by improvements in the level of human knowledge and skills. Where do the new machines and equipment and new production skills come from, which will be paid for by investing this surplus?

B *Transfer of Technology*

Particularly with respect to developing countries, the degree of novelty and level of efficiency of 'new' technology is relative to the technology which it replaces: the appropriate measure of technological advance is restricted to the economy in which the change occurs. In other words, although a machine may have already existed and been in use in one economy for some time, when it is first used in another economy it is 'new' to that production system; thus, regardless of the actual or potential existence of still more efficient technologies in other economies, the replacement of a less efficient by a more efficient technique represents technological advance in that context. A major source of 'new' technology is the transfer of existing technologies from an industrialized country to a developing one. A strategy of technological modernization, therefore, may be based on the importation and adaptation of existing foreign technologies, as well as on the use of technologies invented within the developing country itself.

However, there are a number of factors which limit the effectiveness of this as a strategy for technological advance. First, the technologies of industrialized countries are developed for use in economies with very different factor endowments from those of most developing countries: in addition to different patterns of natural resources and climate, the ratio of labour to capital is commonly higher in developing countries. Although it is sometimes possible to modify or adapt imported technologies to the pattern of factor endowments in a developing country, this adaptation must usually be made in the developing country itself in order to take into account all the relevant conditions of production. However, the owner of proprietary technology in an industralized country often retains control over his technology, merely licensing its use in the developing country, or setting up a subsidiary of his own firm. Even if he is prepared to allow adaptations to be made, he may insist that these be carried out by scientists and technicians from the industrialized country. In this case, the growth of an adaptive and inventive capacity in the developing economy is inhibited.

In practice, the R & D system most responsive to an investment decision—a choice of technique—is that which exists within the same economy or the same political system as the decision-maker, whether an entrepreneur or a planner. If it is true that investment decisions in the interests of development are more likely to be made within the developing country than in an outside economy, then the lack of an indigenous R & D capability may prolong technological dependence and preclude autonomous innovation leading to economic and social development.

There are often further restrictions on a developing country's access to existing technologies. Technology—that is, the knowledge of a particular production process or of the construction of a particular machine—is regarded in many societies as private property, and the inventor's rights to the use of his technology are usually protected *de jure* by patent or *de facto* by secrecy. Even aside from the fact that the individual entrepreneur or the economic planner in a developing country may have only a limited knowledge of the full range of alternative techniques, he may only be able to obtain the use of those technologies he does know about on very restrictive terms, if at all. Furthermore, unless the technology is received as disinterested aid or technical assistance from another country, or unless the transfer is occuring through investment by a foreign or multi-national enterprise, the transfer of proprietary technology will require the developing country to spend scarce reserves of foreign exchange.

In addition to the importation of new capital goods which embody technology, the concept of 'technological advance' also includes the importation of 'know-how'. Here a distinction must be made between 'know-how', defined as the knowledge and skills to operate and maintain existing technologies, and 'know-how' as the capacity to invent and create new technologies and adapt and improve on existing ones. Operational know-how is more or less tied to current production practices and derives partly from on-the-job experience. But because control of imported technology, including the authority to change that technology, remains for the most part outside a developing country, transfers of plant and equipment are not necessarily accompanied by transfers of the know-how necessary for technological innovation.

Technological know-how, in the sense of the capacity to invent new technologies and technological improvements, can also be transferred to a developing country independently of capital-embodied transfers. This can occur when individuals obtain technical and engineering training and scientific education either in institutions in industrialized countries, from which they return home, or in local institutions, which are often modelled upon those of industrialized countries. This training, and the research undertaken by the graduates of these institutions, is often irrelevant to the technological needs of the developing country, particularly when patterns of ownership of technology tend to limit local R & D to a marginal role in any innovation that does occur.

Modern industrial technologies were originally transferred to China in the late nineteenth century and the first half of the twentieth, to foreign-owned (and some Chinese) enterprises in the treaty ports and Manchuria. This industrial base was strengthened and enlarged by substantial imports of Soviet and east European technologies in the 1950s, and by technologies purchased from Japan and western Europe in the 1960s.

Until the mid-1960s, many Chinese engineers and technical personnel were trained abroad: in Europe, North America and Japan before 1950, and in the Soviet Union and eastern and western Europe after the People's Republic was established. The persistence of conventional, imported institutional patterns in education and in R & D in China has been regarded by some Chinese policy-makers as hindering the growth of an indigenous innovative capacity and as prolonging dependence on imported technologies. On the other hand, the problem of 'brain drain' faced by other developing countries—that is, the emigration of highly educated and skilled scientific and technical manpower to the industrialized countries—is not a problem in China because of particular political conditions there.

The items in this section refer to the transfer of foreign technologies to China since 1949. Implicit in most of these is a more positive assessment of the value of technology transfer for development in China than has been regarded as the general experience; however no analysis which would establish a basis for comparison with other developing countries, has been made of the mechanisms by which technology has been transferred to China. The reader is referred to the appendix for items on the transfer of western industrial technologies to China before 1949.

Technology transfer may also occur in the form of technical assistance from a developed to a developing country: when machinery and other forms of technological hardware are either given to a developing country, or sold at a subsidized price, or on special terms of payment, when special training and education programmes are established for personnel from a developing country, or when an industrialized country makes special arrangements to second technicians and engineers to a developing country, then the transfer of technology is a form of technical assistance. Such assistance is not usually offered for disinterested motives, and the recipient country may be exchanging some degree of political or economic independence for the technology received as 'aid'.

China's experience of technical assistance is rather complex. Although a considerable portion of the technology received from the the Soviet Union in 1950s was transferred as 'technical assistance', the Chinese subsequently claimed to have paid in full for this aid. All references to Soviet technical assistance to China therefore, are classified as 'technology transfer' in this bibliography.

### C *Innovation and the Direction of Technological Change*

To sustain economic growth and development, progessively more efficient techniques of production must replace the techniques chosen first. As economic surplus is re-invested in new machinery and equipment and in the education and technical training of the labour force, successive rounds of technological change may occur in either a labour-intensive or a capital-intensive direction. The precedent set in the (labour-scarce) industrialized countries indicates that technological advance has been introduced into the production system *via* capital goods—for example, machinery and equipment—which supplement the productive power and raise the productivity of labour. The existence of such capital-intensive, highly efficient technologies in the industrial economies implies that if technological change continues in a labour-intensive direction in a developing country—as might occur to fulfil certain social or political objectives—some of the surplus output which might otherwise have been obtained with capital-intensive techniques and so have become available for re-investment in still more efficient capital goods, is sacrificed.

Alternatively, new technologies may be invented which are both labour-intensive and highly productive. This would imply the invention of unique technologies, distinct both from traditional, relatively inefficient techniques and from modern, capital-intensive technologies. Or, if social and political circumstances allow wages to be maintained at a low level so that less of the total product is consumed, a correspondingly larger proportion of output is left as surplus for re-investment in either the same, or new, means of production. If the technology in this new round of investment is labour-intensive, more employment is generated, and more of the unemployed or under-employed labour force can be absorbed. If the new technology is capital-intensive, the labour already employed becomes more productive and produces a larger quantity of output, which may be redistributed to the unemployed part of the population by some means other than payment of wages.

It is also open to developing countries to promote mixed economies, in which the labour-intensive, employment-generating technologies are limited to certain specific sectors, such as consumer goods or agricultural production, or small-scale enterprises; while capital-intensive, efficient technologies could be used for the production of capital goods, say, or in some limited sector of the economy in which investable surplus, foreign exchange for the purchase of equipment and materials from abroad, and modern technical and engineering skills are concentrated.

In this section on technology policy in China, the following questions are considered: if the Chinese authorities have succeeded in imposing their particular economic and social objectives on the direction of technological change, and if they have succeeded in making technological advance an instrument for achieving their development goals, has the result been the invention of wholly original, new technologies? Have efficient, labour-intensive, small- and medium-scale technologies been developed to enable industrialization to occur in rural areas where only limited financial and engineering resources are available? Or has the use of comparatively inefficient, labour-intensive technologies, or the maintenance of low wage levels implied that some of the Chinese planners' development goals have been sacrificed or postponed in order to fulfil other objectives? In modern industrial enterprises in China, have original technologies been invented to suit the conditions of production, or have imported technologies and

Chinese-made copies of foreign machines remained the basis of production? What are the implications of 'self-reliance' for technology transfer?

Very little is known about the actual process of innovation in China: the procedure by which new inventions, existing machines and production processes, or imported technologies are selected, tested, modified and improved, adapted from prototype to production scale, introduced into productive use, and diffused to new users. The innovation process seems to differ in the modern industrial sector, in agriculture, and in the small-scale rural industrial enterprises, but the information available from Chinese sources has not yet been fully exploited in the secondary literature.

The literature also reaches contradictory conclusions about the direction in which innovation is occuring in China. One school of thought especially prominent among recent Japanese writers on the subject purports to see the development of a new relationship between society and technology in China, which may or may not imply some degree of originality in that technology. There is, on the other hand a general impression, particularly among engineers and technicians who have studied the 'level' of technology in Chinese industries, that these technologies are neither original nor, in comparison with technologies used in the developed countries, particularly new. Thus the trend of technological change in China may be in the same direction as in other countries—that is, increasingly toward the substitution of capital for labour. The process of innovation in Chinese industry and agriculture, however, may reflect the unique features of economic and social organization in China.

# A Choice of Techniques

27 ISHIKAWA, Shigeru
Choice of technique during the process of socialist industrialization - Comments on Dobb's 'Chinese method.' *Keizai Kenkyu*, 12(3), 267-71, 1961.
Refers to Maurice Dobb, *An essay on economic growth and planning*, London, Routledge & Kegan Paul, p. 46-7, 1960. 'Chinese method' use of primitive local industries which are highly labour-using in supplementation of more technically advanced construction projects. Gross production may increase while employment is extended, although surplus will not increase proportionately as much as the increase in gross production will represent an increase in wages. Unless the additional employment involves no claim on funds of the investment sector, the use of these technologies will reduce the growth rate. A possible off-setting advantage of the 'Chinese method': shorter period of production shortens the time lag, results in increments of investment at earlier dates via the compounding effect.

28 ISHIKAWA, Shigeru
Choice of techniques in mainland China. *The Developing Economies*, Tokyo, Institute of Asian Economic Affairs. Preliminary issue No. 2, 23-56, 1962.
'Experience in China presents an important example revealing the merits and limitations of the economic policy of promoting development of medium- and small-scale firms . . .' I--Size of industrial firms under the first FYP; 'the Chinese tried to follow the normal pattern of development in socialist countries with regard to the scale of establishments and technology.' II--Quantitative evaluation of dualistic development among firms and its revision during the second FYP. III--Reasons for the declining trend in the number of small firms in the case of small-scale iron-works.

29 ISHIKAWA, Shigeru
A note on the choice of technology in China. *The Journal of Development Studies*, 9(1), 161-86, 1972.
Chinese planners' choice between known and available technologies, and between techniques that are immediately applicable to industrial and agricultural production; considered in terms of the planners' basic objectives, their criteria of choice, technical and technological alternatives, and objective conditions which constrain or facilitate the planners' choice; reviewed in the three phases which have characterized Chinese development: the 'Soviet model', 'Great Leap Forward', and '3 and 4 Five Year Plan' phases. The Maximum Growth Criterion and the investment-inducement mechanism.

30 RISKIN, Carl
Local industry and the choice of techniques in the planning of industrial development in mainland China. *Planning for Advanced Skills and Technologies*, Industrial Planning and Programming Series No. 3 (UNIDO). New York, United Nations, p. 171-80, 1969.
'Chinese planners have experimented widely and in many industries

with various alternative methods of producing identical or close substitute products as well as using different techniques in different sectors . . . . A fundamental characteristic of these experiments is their deliberate recourse to technological dualism (or, more accurately, pluralism), that is, their stress on the simultaneous development of techniques of varying degrees of mechanisation and varying combinations of factors of production.'

31 RISKIN, Carl
Small industry and the Chinese model of development. *The China Quarterly*, No. 46, 245-73, 1971.
'This article examines the evolution and implementation of China's policy towards small and medium industry . . . . One of the objectives of this article is to delineate the relation between small industry and the general development problem in China, and thereby to gain some insight into the nature and logic of the particular strategy of development associated with Mao.' Two principles of local industrial policy: 'choice of techniques' function--use of small-scale, relatively labor-intensive techniques; and 'sectoral allocation' function--production of rural producer and consumer goods, esp. for agriculture. Local industrial policy not thoroughly implemented during first FYP. Economic and technological roles of small industry remained similar in the Great Leap, although emphasis on heavy industry and market incentives in the commune industries compromised the 'sectoral allocation' role and 'choice of techniques' function was sabotaged by the disruption of production of rural commodities resulting from appropriation of resources from handicraft production for commune industries. Principle tenets of the small-industry policy today: '(a) to build "small but comprehensive" and relatively self-sufficient industrial systems of dispersed factories operated and controlled by the various localities themselves; (b) to link such systems with the need of agriculture; and thereby (c) to raise labour productivity in agriculture by means of innovation and technological change, including mechanisation.'

32 *CURRENT SCENE* EDITOR
The conflict between Mao Tse-tung and Liu Shao-ch'i over agricultural mechanization in Communist China. 6(17), 1968.

33 DERNBERGER, Robert F.
Economic realities. *Contemporary China*, Ruth Adams (ed.), New York, Vintage Books, p. 125-38, 1966. Also *Bulletin of the Atomic Scientists*, 22(6), 6-10, 1966.
Four major economic problems in China include . . . low technical level and dependence on a foreign assistance. A responsible Chinese government could solve these problems--but the building of heavy industry and a modern military establishment remain priorities of the present government.

34 DERNBERGER, Robert F.
Economic realities and China's political economics. *Bulletin of the Atomic Scientists*, 25(2), 34-42, 1969.
Any meaningful assessment of economic realities in China today must include a discussion of the different economic policies of the

contending factions in the Cultural Revolution on the current state of the economy, and its effect on China's economic future. The anti-Maoist school represents the 'technologist's' approach to economic development, relying on the creation of a technical bureaucracy. Soviet aid to industry.

35 ECKSTEIN, Alexander
Strategy of economic development in Communist China. *American Economic Review Papers and Proceedings*, Vol. 51, 508-26, 1961.
'The central thesis of this paper is that during the first FYP period (1953-57) Chinese Communist policy-makers pursued a Stalinist strategy of economic development with local adaptations. However . . . the Chinese planners evolved a new strategy for the second FYP (1958-62), based on intensive utilization of under-employed labor combined with promotion of technological dualism, as a means of maximizing the rate of economic growth. In effect, then, building on Stalinist foundations, they adopted an essentially Nurkse cum Eckhaus model of economic development.'

36 ECKSTEIN, Alexander
*Communist China's economic growth and foreign trade: Implications for U.S. Policy*. New York, McGraw-Hill, for the Council on Foreign Relations, 1966.
China's economic development theoretically affects its foreign policy in three ways: by affecting its potential to wage war; by affecting its capacity to expand international economic relations; and through the possible appeal of its development model to other underdeveloped areas. This study focuses largely on the second. First, the character and pattern of economic development is analyzed; role of imports in the expansion of industrial capacity; rising export capacity as a means of financing imports for industrialization and the use of trade for political purposes. The technological base; the strategy of development through 'technological dualism'. Soviet assistance to China. Chinese assistance to underdeveloped countries.

37 GRAY, Jack
The economics of Maoism. *Bulletin of the Atomic Scientists*, 25(2), 42-51, 1969. Also Mao's economic thoughts, *Far Eastern Economic Review*, 67(3), 16-18, 1970.
Fundamental tenets of Mao's economic philosophy: 1--emphasis on production increase. 2--material incentives and economic rationality. 3--collective entrepreneurship. Agricultural mechanization; intermediate technology.

38 HINIKER, Paul J. and FARACE, R. Vincent
Approaches to national development in China: 1949-1958. *Economic Development and Cultural Change*, 18(1), Pt. 1, 51-72, 1969.
'We assert that there were two essentially different development models employed in Communist China between 1949 and 1958. The first of these was a Russian model characterised by its emphasis on technology and extensive bureaucratization; the second was essentially a Maoist model involving wide-scale mobilization of the Chinese masses.'

39 MCFARLANE, Bruce
Mao's Game Plan for China's industrial development. *Innovation*, No. 23, 2-13, 1971.
Discussion of Chinese policies on industrial and agricultural technology in the context of policies for long-term economic development and political and ideological goals: 'developing an agrarian socialism and avoiding large-scale labor transfers to the cities.' Distribution of industrial development; emphasis on small- and medium-scale enterprises; plant self-sufficiency as a function of self-reliance and mass participation in production decisions. 'In other words, ideology is influencing technological policy, and at this stage of China's development that ideology may be summed up as the desire to resist excessive technocratic control, and the belief in motivation by non-financial means.' Imitation of foreign machine design; indigenous innovation and design. Obstacles to technological development: unintegrated character of the parts industry prevents emergence of highly skilled technical people; lack of a core of innovation managers; inadequate system of diffusing innovations. '. . . the coming decade is propitious for technology, including use of foreign technology--for acceleration of the rate of economic growth . . . the actual outcome will depend on three factors: first there is the question of the success of the educational and institutional changes sought in the Cultural Revolution . . . . Second, in the growth factors, there is considerable scope for raising productivity in agriculture . . . . The third factor is an ability to purchase foreign technology.' 'I have a distinct feeling that one of the advantages of the Chinese approach so far is that it regards economic growth as more than a mere function of physical investment and technological progress . . . . The Chinese contend that the human factor--including motivation--is a vital part of the equation leading to economic development.'

40 OVDIYENKO, I. Kh.
The new geography of industry of China, *Geograf v. Shkole*, No. 6, p. 28-41, 1959. Also *Soviet Geography* (American Geographical Society), 1(4), p. 63-78, 1961.
'The chief task of the socialist industrialization of China consists in creating in about three five-year periods a basically integrated industrial system in which heavy industry will predominate.' Large modern industrial enterprises are the 'backbone' of production but still cannot satisfy the requirements of the expanding national economy. Thus, small and medium-sized enterprises, which require small investment, can be put into operation quickly and use local materials and waste products, are being developed, especially in rural regions. Distribution of new industrial construction near to sources of raw materials and fuel to markets--in interior regions. More than half of 921 major industrial construction projects begun during first FYP are located in the interior. Review of coal industry; petroleum industry; electric power industry; iron and steel industry; machine building industry; chemical industry; textile industry.

41 SIGURDSON, Jon
Teknologisk utbveckling i Kina (China's technological development). *Internasjonal politikk*, No. 2, 1970.
The Great Leap Forward was a recognition by the Chinese that they

had to look for their own solutions to realise the quickest possible economic development. Chinese leaders have divided over the question of technological self-reliance during the Cultural Revolution. Development of intermediate technology may preclude achievement of catching up with advanced science within 20 to 30 years.

42 YEH, K.C.
Soviet and Communist Chinese industrialization strategies.
*RAND P-3150*, May 1965.
Comparison of Soviet and Chinese approaches to problems of allocation of resources to investment; priority of industry in the allocation of capital investment; choice of techniques and scale in industrial production during their respective first FYPs.

43 ZAUBERMAN, Alfred
Soviet and Chinese strategy for economic growth. *International Affairs*, 38(3), 339-52, 1962.
'The purpose of this article is to sketch (the) broad lines (of the Soviet model for economic growth) and to enquire where and how far the Chinese strategy deviated from the Soviet prototype.' '. . . on the whole it is the producer goods, or the heavy, or capital-intensive, industries that are the principal carriers of technological progress . . . . As a rule, in a modern industry, variation limits of capital-to-labour proportions are becoming increasingly narrow, hence the dilemma facing promoters of initial industrialization is . . . either to condemn the industry to technological backwardness at its birth, or to deepen structural under-employment.' Chinese practice follows Soviet policy for dealing with the conflic between industrialization and factor endowment; and 'consciously tries to preserve and vastly to expand small-scale and cottage-type sectors.'

See also the following entries:

Technology and Economic Growth. 1, 4, 6, 12, 13, 14, 17, 20, 22.
Technology Policy:
  Transfer of Technology. 48.
  Innovation and the Direction of Technological Change. 60, 62, 66, 85, 93, 94, 95, 99, 103, 109
Policy for Science. 150, 175.
Agricultural Technology. 482.
Industrial Technology:
  Chemical Industry. 522, 526.
  Petroleum Industry. 533, 540.
  Machinery Industry. 576.
  Basic Metals Industries. 607, 609.
  Mining Industry. 625, 626.
Engineering. 635.
Transportation and Communication. 646.
Trade. 658, 659, 661, 664, 665, 666.
Medicine and Public Health. 685, 686, 687, 688, 690, 696, 698, 699, 700, 703, 706, 714, 715, 716, 721, 725, 731, 733.

## B Transfer of Technology

44 CHENG, Chu-Yuan
*Economic relations between Peking and Moscow: 1949-63*. New York, Praeger, 1964.
'The purpose of this paper is to show that the pattern of Communist China's economic development and development policy itself was determined, to a significant extent, by developments in Communist China's foreign trade and capital movements.' In the early 1950s, Soviet loans, supplies of machinery and equipment and technicians made possible high rates of growth and creation of a capital-intensive heavy industry; change in capital flow after 1955 led to initiation of the Great Leap Forward, which aimed to establish with native technology a labor-intensive industrial sector . . . future developments of the heavy industrial sector will continue to depend upon Communist China's foreign trade and capital movements and the import of foreign technology.

45 CHIN, Calvin Suey Keu
*A study of Chinese dependence upon the Soviet Union for economic development as a factor in Communist China's foreign policy*. Communist China Problem Research Series EC22. Hong Kong, Union Research Institute, 1967.
'The purpose of this study is to analyze . . . how (China's) longing for industrialization assists Soviet pressure in maintaining an economically dependent China . . . . Russia is providing China with the capital goods and the technical assistance she desperately needs in economic development.' Soviet technical assistance helps to bind China for the present to one system, for Soviet blueprints, specifications, and techniques are widespread and in use throughout China. V--Soviet Aid: the nucleus. Pt. III--Technical assistance. Pt. IV--China's student program. VI--Influences of Soviet aid: maintenance of a dependent China. Pt. III--The role of technical assistance.

46 *CHINA NEWS ANALYSIS*
Rebuilding political cooperation. No. 179, 1957.
Soviet experts p. 5-6.

47 *CURRENT SCENE* EDITOR
Decision for an 'upsurge': Some impressions on Peking's approach to economic problems. *Current Scene*, 3(17), 1965.
P. 8. Shortage of manpower in the applied sciences. Peking's policy is to use native resources where possible, but also to use foreign technology and equipment, train own specialists, import and copy technology and model plants, resist accepting direct technical assistance.

48 DERNBERGER, Robert
The relationship between foreign trade, innovation, and economic growth in Communist China. *China in Crisis*, Ho, Ping-ti and Tsou, Tang (eds.), Vol. 1, 739-52. The University of Chicago Press, 1968.
Four integral aspects of Sino-Soviet economic relations, 1949-63: Soviet assistance to China's industrialization program, with

particular emphasis on the supply of equipment and technical aid; Sino-Soviet commercial relations; nature and amount of Soviet financial aid; prospects for future developments. Chap. 3--Soviet assistance in the economic reconstruction of Communist China: the scope and nature of Soviet aid; Soviet supply of equipment and machinery; Soviet technical assistance; Soviet experts working in China; exchange of technical data and blueprints; training of Chinese specialists and technicians in the Soviet Union; training technicians and skilled workers in China; the modified pattern of Soviet aid.

49 GOODSTADT, Leo F.
Leaping backwards. *Far Eastern Economic Review*, 56(45), 303, 1969.
Confusion in Chinese policy on import of technology.

50 INSTITUT FÜR ASIENKUNDE (Hamburg)
*Die wirtschaftliche verflechtung der Volksrepublik China mit der Sowjetunion*. Frankfurt am Main, Alfred Metzner Verlag, 1959.
Economic relations of the People's Republic of China with the Soviet Union. 4--Soviet aid to the industrialization of China. (2) Scientific-technical cooperation.

51 KLATT, Werner (ed.)
*The Chinese Model*. Hong Kong, Hong Kong University Press, 1965.
8--Sino-Soviet economic relations: Chinese dependence on Soviet aid. Includes sections on Soviet technical aid.

52 KLEIN, Sidney
Sino-Soviet economic relations, 1949-62: a Sinologist's sketch. *Current Scene*, 2(15), 1963.
Russian-provided industrial machinery and equipment provided the major substance of the Chinese Communist industrialization effort; as Soviet aid to Peking ebbed, so did the rate of industrialization in China.

53 PARKER, David
Travelling Chinese. *Far Eastern Economic Review*, 74(42), 57, 1971.
Visit by a Chinese team of telecommunication experts to France, Sept 1971. '. . . this practice of sending "study groups" of engineers and other specialists to the more technologically advanced nations around the world has seemingly become an article of faith with the Chinese and is an established feature of modern China's scientific and industrial effort.'

54 SCHUMAN, Julian
Technology interest. *Far Eastern Economic Review*, 78(52), 31-2, 1972.
The main Chinese emphasis in the field of imports is on technology and equipment with a high technical content. Competition for the China market by the industrialized nations is now directed towards this need. The China market will not be cornered by any one nation. Recent trade developments between China and Britain.

55 WILSON, Dick
China's economic situation. *Bulletin of the Atomic Scientists*, 23(9), 3-8, 1967.
Section on foreign aid and self-reliance.

56 WILSON, Dick
China's industrial prospects. *Far Eastern Economic Review*, 45(7), 272-3, 1964.
The development of Chinese industry will depend on a reasonable number of Chinese engineers and scientists remaining in the full swim of the world's technological progress. Without one industrialized power politically committed to helping China, recruitment of foreign technicians is difficult. The Japanese are an obvious choice to replace Russian technical assistance.

57 YAMANOUCHI, Kazuo
Communist China's dependence upon the Soviet Union in her economic development. *Ajia Keizai*, 1(4), 13-22, 1960.
Section II--Contributions of technical aid from Soviet sources. Statistics.

See also the following entries:

Technology Policy:
  Technology and Economic Growth. 8, 10, 21, 24, 26.
  Choice of Techniques. 41.
  Innovation and The Direction of Technological Change. 79, 87, 110.
Policy for Science. 138, 152, 175, 188.
Research and Development:
  Electronics. 271.
  Military/Space Exploration and Travel/Nuclear Energy. (see also Technical Assistance) 324, 335.
Industrial Technology:
  Electric Power Industry. 497, 498.
  Electronics Industry. 507, 513, 516, 518.
  Chemical Industry. 523, 527, 528, 531, 532.
  Petroleum Industry. 533, 536, 537, 538, 539, 546, 551, 556, 557, 558, 559, 562.
  Textile Industry. 571, 574.
  Machinery Industry. 575, 576, 579, 590, 592.
  Transport Equipment Industry. 596, 597.
  Basic Metals Industries. 600, 601, 610.
  Mining Industry. 624, 627, 628.
Trade. 660, 661, 662, 664, 665, 666, 668, 669.

## C Innovation and the Direction of Technological Change

58 AKABANE, Nobuhisa
Consideration of methodology in the analysis of technological levels. *Long-term prospects for the Chinese economy*, Shigeru Ishikawa (ed.), Pt. III, Chap. III, No. 119, Tokyo, Asian Economics Research Institute, 1967.

59 ANDORS, Stephen
Revolution and modernization: man and machine in industrializing society, the Chinese case. *America's Asia: dissenting essays on Asian-American relations*, Friedman, Edward and Selden, Mark (eds.), New York, Random House, 1969. Also New York, Vintage Books, p. 393-438, 1971.
'This study of factory management . . . is an attempt to confront (another) problem of the contemporary age: how man can organize complex human interaction around the technology he uses to control his environment . . . how alienation of man from the products and processes of production can be overcome . . .' Patterns of factory management and other technological innovation and technical processes: the Great Leap Forward; 1960-63; 1963-66; the Cultural Revolution. 'The questions raised by factory management in China . . . touch the basic question of human identity.'

60 BERGER, Roland
Self-reliance, past and present. *Eastern Horizon*, 9(3), 8-24, 1970.
'Self-reliance' prior to 1949; in the early years of the PRC, when the emphasis on self-reliance 'receded' in the face of an assumption that Soviet technical and organizational methods would suit Chinese conditions . . .', movement for technical innovation during the Great Leap Forward; self-reliance during the Cultural Revolution: launching of an earth satellite; barefoot doctors; agriculture; small scale industry. Economic planning. China aims to develop organic communities with both an industrial and agricultural base, grouped into medium-sized and larger social and economic areas, rather than big cities and conurbations.

61 CHAO, Kang
*The rate and pattern of industrial growth in Communist China.* Ann Arbor, University of Michigan Press, 1965.
An attempt to construct an independent index of the industrial output produced by Communist China between 1949-1959 primarily based on the officially published data of physical output. p. 57-75. Technical coefficients of industrial production.

62 CHEN, Jack
Taking off on a tricycle. *Far Eastern Economic Review*, 73(36), 26-8, 1971.
China's strategy for progress is based on more or less self-sufficient industrial zones, formed by the coordination of large, medium and small plants. This policy was first articulated by Mao in April 1956. The author's observations of this principle in action in Shanghai.

63 *CHINA NEWS ANALYSIS*
Economic growth and blunders.
*Statistical Workers Bulletin*, No. 199, P. 4, 1957.
Technical level of industrial production.

64 *CIENCIA Y TECNICA EN EL MUNDO*
El Desarrollo Cientifico e Industrial en China Continental. 23(396), 445-66, 1971.
1--The evolution of science and technology. 2--The innovation process. 3--Recent progress in Chinese industry, including technical evolution, by sector. Conclusions: changes in the structure of China's industry and mass participation of its citizens in the campaign for technical innovation is beginning to bear fruits.

65 *CURRENT SCENE* EDITOR
Industrial development in China: A return to decentralization.
*Current Scene*, 6(22), 1968.
The controversy between vertical, specialized industrial organization and decentralized, fully integrated industrial organization. 'Self-reliance' and 'technical innovation' vs. 'standardization', 'specialization' and 'quality control'.

66 DEAN, Genevieve C.
Innovation in a choice of techniques context: The Chinese experience 1958-1970. *Bulletin of the Institute of Development Studies*, 4(2/3), 39-48, 1972.
Consistency of policies since the Great Leap Forward for technological advance in state-owned industries using imported modern, large-scale, capital-intensive technologies and in the agricultural and consumer goods sectors which use traditional, small scale, labor-intensive technologies.

67 *FAR EAST TRADE AND DEVELOPMENT*
How China is managing. Annual China Review, 27(10), 438-41, 1972.
Agriculture; irrigation; power. Transport. Iron, steel and fuels (coal, petroleum). Electronics. Railways. Heavy machinery. Description of technology in Shanghai Industrial Exhibition.

68 HAMBRAEUS, Gunnar
Experience from China just now. *Teknik och Industri i Kina*. Stockholm, IVA (Ingenjörsvetenskapakademien), Report 44, 1972.

69 HAN, Suyin
*China in the year 2001*. Harmondsworth, Penguin Books, 1970.
The author attempts 'to explain the Chinese way of grappling with their own situation and problems.' 4--Trade, population, research, national economy, factors of inhibition. Science: Development and research. China has not only conducted the steps necessary for the first industrial revolution, but has also undertaken the second revolution necessary because of advances in automation and in the use of atomic energy; two aspects to the research program are the high level concentration of the best brains and talent on certain projects, the second, the most widespread diffusion, communication and facilities for creating a pool of technological personnel of all types at worker-peasant level. Scientific

achievements. Agricultural and industrial research. Nuclear development. 3--Industry: Designing a new heaven and earth. Russian technical aid; technological innovation; technological manpower and organization of technical matters.

70 HERON, Antoine
Gestion revolutionnaire dans les enterprises. *La Nouvelle Chine*, No. 6, 23-7, 1971.
Revolutionary Management of Enterprises. 1--Proletarian power in the enterprise. 2--Specific objectives and mass movements. 3--Methods of labor: Organization of the enterprise; The technical system; Participation of workers in technological research.

71 HIRAMATSU, Shigeo
China's Socialist construction and 'self-reliance'. *The Journal of National Defense*, 10(4), 7-54, Pts. 1, 2, 1972.

72 HOFFMAN, Charles
*Work incentive practices and policies in the People's Republic of China, 1953-65.* Albany, State University of New York Press, 1967.
The promotion system stimulates workers to increase their labor input and improve their skills, and furnishes growing numbers of technicians and professional personnel for industry. Pay scales for professors, higher level scientific personnel, technical workers. Incentives for invention, innovation and scientific research for industrialization. Mass technical demonstrations as incentive.

73 HOSHINO, Yoshiro
*Basic problems of technological innovation.* 2nd ed. Tokyo, Keizo Bookstore, 1969.
The technological innovation movement occuring in China. X--Technological innovation movements before the general line of Socialism (The Great Leap Forward): The direction of Soviet technologists' technological innovation movement; The advanced producers' movement; The problem of scientists and technicians. XI--The technological innovation movement under the general line of Socialism: Technological innovation movement arising from agricultural collectivization; Significance for technology of industrial construction on two legs; Technological innovation movement by means of the combination of three; The problem of thought reform of intellectuals. XII--Divergence between the Chinese line and the Soviet line: The period of difficulties of the general line of socialism; Advocacy of the movement for three great revolutions; The Great Proletarian Cultural Revolution and the revolution in education; The Great Proletarian Cultural Revolution and scientists and technicians.

74 HOSHINO, Yoshiro
China's science and technology and Japan's science and technology. *Chugoku*, No. 104, 6-31, 1972.
Comparison of trends of technological development in China and Japan. The stagnation of technological innovation. Modern science and technology cannot solve problems of public harm. Urbanization

of agricultural villages, ruralization of cities. Decline of farming and fishing villages with industry. Communist unrewarded labor and science and technology's philosophy.

75 HOSHINO, Yoshiro
China's technological line in the Great Proletarian Cultural Revolution. *Metals Magazine*, 64-78, June 1971.
Technical conditions peculiar to socialism. Basic starting point of China's cultural development. The contradiction between technical specialization and the masses. Fundamental change in engineering colleges. Criticism of the ideology in engineering textbooks. The problem of the reflection of classes in technology. New technological development with agriculture as the base and industry as the leading factor.

76 HOWE, Christopher
Problems, performance and prospects of the Chinese economy. *The World Today*, 23(12), 529-34, 1967.
General discussion of the problems of allocating resources for investment; choice of economic institutions and incentives; generating a flow of scientific and technical innovation and incorporating innovations into the productive process.

77 ISHIKAWA, Shigeru (ed.)
*Long-term prospects for the Chinese economy, Part II*. Tokyo, Asian Economics Research Institute, No. 102, 1966.
Discussion meeting: Special technological features of Chinese medium- and small-scale industry.

78 KOJIMA, Reiitsu
China's indigenous technology. *Technology and People*, Pt. 1, 57-61 Spring 1972; Pt. 2, 57-66, Summer 1972.
1--Superstitions of the Chinese masses and 'science'. 2--Western European science and technology and revolutionary thought. 3--The Great Leap Movement, the theory and two kinds of superstitions, and the transformation of people: The first FYP; Eradication of blind belief in foreign science and technology; The core of eradicating two kinds of superstition; New idea of technology--What is the idea of indigenous technology; Two trends in invention. 4--The stage after the Cultural Revolution.

79 KOJIMA, Reiitsu
The direction of technology in China. *Metals Magazine*, 38(20), 15-18, 1968.
Chinese economic and technical organization. The foreign technology model: first Five Year Plan. Origin of the idea of indigenous technology and the Great Leap Forward. The imitation of Foreign Technology Party and those responsible for indigenous technology. Formation of the two schools into opposing groups.

80 KOJIMA, Reiitsu
On an 'Integrated National Democratic Economy'. *Ajia Keizai*, 6(9), 4-16, 1965.
I--Contents of 'Integrated National Democratic Economy'. The actuality of the 'Integrated Industrial System' seen from the machine

tools industry (comparison of technical specifications of Chinese and counterpart tools). II--The meaning of the direction of China's reconstruction for economic development. Potential for technological development of construction machines. The division of socialism and China's technological development (including North Korea).

81 KOJIMA, Reiitsu
On intelligent technology: indigenous thought supports China's economic base. *Chugoku,* No. 105, 6-25, 1972.
The development of Capitalism and the loss of intelligence. Communalization depends on the collectivization of production technology. Definition of creation--indigenous thought. Liberated technology is profitable technology.

82 KOJIMA, Reiitsu
Mao Tse-tung's thought on science and technology. Science and Technology in Modern China, *The Asahi Asia Review,* 3(1), 10-17, 1972.

83 KOJIMA, Reiitsu
Reappraisal of the Great Leap Forward Policy: With special reference to the industrialization of rural economy. *Ajia Keizai,* 8(12), 3-37, 1967.

84 KOJIMA, Reiitsu
'Self-sustaining national economy' in mainland China. *The developing Economies,* 5(1), 50-67, 1967.
'This article looks into the ideology and actual situation behind these slogans (of "the construction of a self-sustained national economy" and an "independent, balanced, modern economy") and handles the applicability of these Chinese policies to the developing nations which are facing the same kind of problems as China.' 'China's experience tells us that . . . even (Soviet aid to heavy industry) was not sufficient to solve the specific economic problems inherent in China.'

85 KRAAR, Louis
I have seen China--and they work. *Fortune,* 111-17, 210, 212, August 1972.
Observations from the author's visit to China 'a few months ago'. Description of technology in county industries, household factories and cooperatives in Canton, Kwangchow Bicycle Plant. 'At every level, China mobilizes its abundant resources of labor . . . to compensate for a scarcity of capital.' Factory management, technical organization at Kwangchow Heavy Machine Tool Plant. Production of agricultural equipment and machinery in Shunte County, Kwangtung.

86 LAURENT, Philippe
Principes ideologiques de la politique industrielle. *Projet,* 1187-202, December 1971.
Scientific research. Development of local small-scale industry. Multi-purpose utilization.

87 LEE, Rensselaer W. III
Ideology and technical innovation in Chinese industry, 1949-1971. *Asian Survey*, 12(8), 647-61, 1972.
'Possibly the most significant expression of the Chinese participatory style is in the technical sphere, where the Communist leadership has sought to promote the application of mass creative intelligence to improving the nation's productive capacity.' 'Concepts of mass technical experimentation and innovating derive from a complex ideology which has adapted elements of Marxist thought . . . to China's specific historical circumstance as a latecomer to industrialization.' 'The equalitarian aspects of the ideology have acquired special validity as a result of China's need to borrow technology from abroad. Technological borrowing . . . fostered the emergence of a new technocratic-managerial elite, but also rendered this elite peculiarly vulnerable to ideological attack.' Innovation in the early 1950s. The Great Leap Forward. The return to politics: the designing revolution to the Cultural Revolution. Technical themes in the Cultural Revolution. Conclusion: 'mass innovation may have performed some important rationalizing functions in Chinese industry by subjecting foreign technological inputs to broadly based scrutiny and criticism. Its social costs, however, are also apparent, particularly its impact upon the morale of professional elites.'

88 MACDOUGALL, Colina
China's industrial upsurge. *Far Eastern Economic Review*, 49(10), 421-3, 1965.
To expand technical expertise in line with their industrial development the Chinese promote 'technical innovations' to replace the trained talent which China does not have either at home or from abroad. China now claims to have a corps of designers who can design large modern plants and mines.

89 MACDOUGALL, Colina
Learn from Shanghai. *Far Eastern Economic Review*, 43(1), 46-7, 1964.
Labour sent to Shanghai factories to learn latest techniques.

90 MACDOUGALL, Colina
The struggle to come. *Far Eastern Economic Review*, 68(23), 59-61, 1970.
Survey of Chinese industry; promotion of local industry.

91 MATSUMOTO, Hiroichi
Technological innovation in Communist China. *Ekafe Tsushin (ECAFE Bulletin)*, No. 248, 1960. Trans. *JPRS*, 7720, Feb 1961.
The nation-wide movement for technological innovation and reform is a part of the general line for Socialist construction: origins of the movement; accomplishments in heavy industry, mining and agriculture; factors contributing to success; comparison of patterns of innovation in China with the Soviet Union, capitalistic nations, Japan. Conclusion: two components of the movement for technological innovation are adoption of latest techniques and promotion of creativity and invention. Present state of technological innovation movement. Measurement of the contribution

of the innovation movement to improvement of China's technological standard. Detailed study of technology in Chinese coal industry. Based on original Chinese sources, including report by Liu Shao-ch'i to 8th Party Congress, 2nd session, May 1958; Kan Kuang-yuan in *Hsueh-hsi;* Po I-po in *Red Flag* No. 10, 1960; *One Great Decade.*

92 MCARTHUR, H. Russell
Technology and technical education in China. *Peace Research Review,* 1(6), 1967.
Contents include background; the agricultural base; transportation; mining and metallurgy; petroleum; hydroelectric power; machine tools; heavy industry; industrial production on the communes; the nuclear program; evaluation of technology; technical education; higher education; educational theory; practice of technical education; research; part-time education; worker research; conclusion: the Cultural Revolution.

93 MCFARLANE, Bruce
Letter to the editor. *Scientific American,* 219(2), 6-7, 1968.
Commentary on Genko Uchida, 'technology in China' (entry No. 110) after the writer's trip to China. 'A number of developments in China since his article appeared and additional Chinese comments on the issues he raises make further remarks on his article worthwhile.' Chinese economic strategy of self-reliance and medium-scale technology; local self-reliance; shortage of engineering and designing expertise dictates concentration of investment in certain sectors (McFarlane: defense and agriculture) and postponement in others (McFarlane: industry). The Chinese planning system allows various levels of productivity within the same industry; 'moral incentives' to enforce managerial efficiency . . . . Comments on specific points about technology made by Uchida.

94 MENDELSSOHN, Kurt
China's little leaps. *The Nation,* 519-21, Jun 1961.
Observations made on the author's 1960 visit to China. Technology in commune industries: low productivity is less important than the fact of production by otherwise unemployed labor; small-scale production will probably be a passing phase of short duration in industrial cities, where large industrial plans are underway, but may become a permanent form of winter employment in agricultural communes. Training of technical manpower: universities pursue 'scholarly study and research', specialized institutes provide intensive training in a fairly narrow field for technicians, technical universities are intermediate between purely academic and specialized courses. Competition by employers for university graduates.

95 MENDELSSOHN, Kurt
The impact of technology. *The Listener,* 257-9, Feb 1961.
Despite low productivity, backyard industry in China fulfills a significant role: by utilizing the enthusiasm of the Chinese people, stimulating inventiveness and ingenuity; by providing technical training; by capitalizing on the 'native skill and

manual dexterity' of the Chinese peasants. Higher technical education: all technological studies have been placed in specialized institutions which conduct research and provide intensive training in a limited technological subject--in order to proceed 'unhampered by a necessarily slow academic development'; basic scientific problems are taught and studied in the science faculties of the universities; applied technological learning with a broader basis is conducted in technical universities.

96 *NEW SCIENTIST*
Mao's ideology in the factories. 'Notes on the News', 41(631), 56-7, 1969.
Brief review of article in *Industrial Ceylon* on worker innovations in China.

97 *PROGRES SCIENTIFIQUE*
'Chine: Developpements scientifiques et industriels en Chine'. 61-79, Sep 1970.
I--Scientific and technical evolution. A--New Structures: general organization of education; primary and secondary education; scientific and technical research; medical circumstances; agricultural circumstances. B--Important results obtained: physics, atomic and space research; chemistry; geology and earth sciences; medicine; agriculture. II--The pursuit of innovation; stages of the campaign for innovations. A--The birth of ideas. B--The men; behavior of technical cadres in posts of responsibility; young technicians and old workers. C--The results. D--Preparation for the future. E--Conclusion. III--Recent progress in Chinese industry. A--Reform of the structure of enterprises. B--Technical research: the situation on the eve of the Cultural Revolution; the reorganization of technical research: 1968-1969; the first results: 1970. C--Evolution of technology, by sector: mining and geology; metallurgy; machine building; electrical construction; electronics, calculators and precision implements; chemicals; petroleum; textiles; civil engineering.

98 RICHMAN, Barry M.
*Industrial society in Communist China; a first hand study of Chinese economic development and management*. New York, Random House, 1969.
A multidisciplinary approach to industrial management and economic development; central thesis is that effective and efficient enterprise management is the key to industrial progress and general economic development. Data derived from two month visit to China in 1966. Pt. I--General conceptual framework. Pt. II--Environment--education, including vocational and technical training, higher education, coordination of education with industrial requirements and manpower utilization; socio-cultural factors, including attitude toward scientific method, risk-taking and change. Pt. III--Structure, operations and performance, including product development and technology, survey of managerial know-how and technology in Chinese industrial enterprises. Conclusion: very few poor countries have done as well in economic growth or industrialization as China has since 1950, but China's fulfillment of her potential depends chiefly on whether

ideological extremism or managerial, technical and economic rationality prevails.

99 SANDERS, Sol and KOLB, John
China drives toward her own technology. *Product Engineering*, Vol. 30, 24-8, 1959.
Description of technology and engineering in China. 'Walking on two legs' in technology: scientific and technical aid from the USSR, Academia Sinica, the Twelve Year Plan for Science and Technology, lack of trained personnel, higher education, political constraints on scientists and engineers; and use of traditional techniques, substitution for scarce materials. 'Throughout history, China has repeatedly absorbed conquerors and made their culture a part of her own. Therefore we can look forward to a technology molded by the "walking on two legs" concept.'

100 SATO, Masumi
China's technological development in the last ten years. *Long-term prospects for the Chinese economy*, Shigeru Ishikawa (ed.), Pt. III, Chap. V. Tokyo, Asian Economics Research Institute, No. 119, 1967.

101 SATO, Masumi
China's technological level and pattern of development. *Long-term prospects for the Chinese economy*, Shigeru Ishikawa (ed.), Pt. I, Chap. IV. Tokyo, Asian Economics Research Institute, No. 76, 1964.

102 SATO, Masumi
Chinese medium and small scale industry seen from the technological angle. *Long-term prospects for the Chinese economy*, Shigeru Ishikawa (ed.), Pt. II, Chap. VI. Tokyo, Asian Economics Research Institute, No. 102, 1966.

103 SATO, Masumi
The realities and technological trends of China's heavy and chemical industries. *Continental Affairs Series*. Tokyo, No. 94, 1967. Trans. in *Chinese Economic Studies*. New York, No. 3, 161-93, 1971.
Review of technological levels and development in Chinese industry by a Japanese engineer, who, as a member of the Chinese Affairs Research Committee of the Asian Economic Research Institute under MITI, has been systematically studying the trends of China's industry for the past four or five years. Real strength of China's industry: highly modernized industries are at the level of about 20% of those in Japan; overall level, including medium and small enterprises and industries, is close to that of Japan. Characteristics of China's industry: coexistence of an extremely advanced area with an extremely backward one . . . because lack of capital limits investment; but the technologically advanced level should be the basis for estimating future development. Trends of development in individual industries: iron and steel: review of technological capabilities; machine industry: Chinese development of machine-tool industry was 'stunning' after suspension of trade with Soviet Union. High

degree of precision of Chinese machine tools, but durability and retention of precision are probably still backward. China is advanced in theory and designing technique. Assessment of Chinese machines produced by machines: generators, petroleum equipment, motor vehicles, electronics, transistors, computers; Chemical industry: fertilizers, insecticides and antibiotics, petrochemicals, synthetic fibres, plastics, synthetic resins; Textile industry: cotton, spinning technology. Future problems: '. . . it is important for China to obtain information on European, U.S. or Japanese technology. I believe this is much more important than the question of importing facilities.'

104 *SCIENCE MAINICHI*
China: Indigenous thought and reality. 31(1), 39-68, 1971.
Keiji Yamada, Reiitsu Kojima, Yuzoe Kato (panel discussion): modern victorious China (economic development), p. 39-49. Table: China's technical development over 20 years, p. 50-5; Kiyoshi Yabuuchi--Projection of the future from the past, p. 56-8; Nobuhisa Akabane--Two kinds of differences in technology, p. 59-63; Ako Akagi--Cybernetics and 'native', p. 64-8.

105 SIGURDSON, Jon
Commentary on industrial policy and technological development in China. *Teknik och Industri i Kina,* Stockholm, IVA (Ingenjörsvetenskapakademien), Report 44, 1972.

106 SIGURDSON, Jon
Factories in the fields. *China Now,* No. 23, 5-6, 1972.
Descriptive analysis of county-level industrial network geared to serve agriculture. Social implications of diversification of rural economy. Indigenous technology and self-reliance in rural industries; adaptation of technology from the cities.

107 SIGURDSON, Jon
Rural industry--A traveller's view. *The China Quarterly,* No. 50, 315-32, 1972.
Observations from the author's visit in December 1971 to some 20 small industrial enterprises in two counties, in Hopei and Honan. Agricultural mechanization. Building local industry: the pilot plant approach; diversification; interdependence of enterprises; indigenous equipment and innovations; management. Three objectives of rural industrialization: acceleration of development of industry; geographical distribution of industry; promotion of technology and research suited to China's conditions.

108 SUGA, Sakae
The level of China's industrial technology. *Continental Affairs Series,* No. 93, Tokyo, 1967. Trans. in *Chinese Economic Studies,* No. 3, 132-60, New York, 1971.
General review of the history of Chinese economic development strategies since 1949, emphasizing technology. 'Therefore, China's goal can be summed up as follows: the goal of socialist construction is industrialization; in order to attain that goal, a modern, independent system has to be established; this, in turn, calls for carrying on construction on the basis of

self-reliance and technical reform or, in other words, founding production on the most up-to-date technology.' Soviet economic and technical assistance to China. The 'scientific experiment movement' as part of industrialization and the policy of self-reliance during the policy of adjustment in the 1960s. Detailed discussion of industrial development in various sectors, based on the literature and at Japan Industry Exhibition and the Chinese Trade Fair. Introduction of technology and technical reform: Soviet technical assistance to 1956; self-reliance and technical reform, starting with the second FYP. Development of indigenous scientific research capabilities: scientific manpower; organization of scientific institutions.

109 SUGANUMA, Masahisa, KOJIMA, Reiitsu and YAMANOUCHI, Kazuo
The direction of development of the Chinese Socialist economy: study of China's local small-scale industry. *Monthly Bulletin of China Research*, No. 278. Tokyo, China•Research Institute, 1971.
I--The GPCR and economic development. II--The development of China's local small-scale industries. III--Local small-scale industries and China's Socialist economy. Discussion: Socialist construction and local small-scale industries.

110 UCHIDA, Genko
Technology in China. *Scientific American*, 215(5), 37-45, 1966. Also *China in Revolution*, Vera Simone (ed.), 363-75. New York, Fawcett, 1968. Also *Chinas Wirtschaft Holt auf*. Eine analyse der industriellen branchen, der technik und der organisation industriekurier, p. 10-11, 13 May 1967; p. 7-8, 20 May 1967.
Review of China's stated technical goals and policies, problems and extent of her advancement in key industries. China is more a 'developing' country than a communistic one. China now seeks to develop a technology of design by seeking technical assistance and knowledge from non-Communist countries. The unorganized structure of China's industry and technology is an obstacle to the development of specialized skills and efficient production. Increasing emphasis on mechanization and automation. Growth of technology in iron and steel, chemical and machine industries. China is ten to fifteen years behind the technologies of modern industrialized countries. See also Bruce McFarlane, entry No. 93.

111 UCHIDA, Genko
*White paper on present day Chinese technology*.
Source ?
Effect of the Cultural Revolution. China's present and future as seen by the technologists. The 'scientific documentation' approach; level of technology before the Cultural Revolution; occurrences during the Cultural Revolution; the iron and steel industry, the chemical industry; machine tools industry; support for technicians; the meaning of self-reliance; characteristics of technological development in China; the rush of the Cultural Revolution; possibilities of Sino-U.S. rapprochement.

112 UNGER, Jonathan
Mao's million amateur technicians. *Far Eastern Economic Review*, 72(14), 115-18, 1971.
Publicity accorded model counties, factories, communes and

production brigades in the Chinese press 'provide glimpses of the means by which the Chinese disperse technical knowledge and co-ordinate their innumerable industrial schemes into a coherent whole'. Themes include 'initiative and technical leadership by unlettered workers; "self-reliance" and thrift by model factories and remarkable technical advances by small under-capitalised plants.' Case studies of worker innovations drawn from the Chinese press.

113 WU, Yuan-li
*The economic potential of Communist China.* Stanford, Defense Analysis Center, Stanford Research Institute, 1963.
Vol. I, Pt. 1, Chap. 7--The industrial base and the state of technology. Vol. 2, Appendix H--Research and development: 'There appears to be a considerable gap between Communist China and the more industrially advanced nations of the world in the fields of science and technology.' 'Efforts to date have been along particular, specialized lines. They have created a working basis for industrial development, considerable reliance having been placed on technical help from Soviet Russia.' Vol. 3 --Reappraisal, 1962-1970. Chap. 3--Agriculture, 1962 and beyond: mechanization, chemicalization.

114 YAMADA, Keiji
Labor, technology and human beings. Pt. 2, Chap. 2, Socialism and human recovery. *China from now on*, Hota Zenbei (ed.), 2nd ed. Tokyo, Chikuma Bookstore, 1972.
Technology in the structure of social relations in Chinese industry.

See also the following entries:

Technology Policy:
  Technology and Economic Growth. 7, 11, 20, 22, 23.
  Choice of Techniques. 39.
Policy for Science. 150, 176, 184, 188.
Industrial Technology:
  Electronics Industry. 508.
  Chemical Industry. 531.

# SECTION III

## *Science Policy*

The definition of 'development' that is used here specifies economic growth which is brought about by technological changes occuring in response to certain economic and social demands arising within the developing country itself. A premise of science policy studies holds that scientific research is a systematic source of the new inventions needed to meet technological 'demand', and that the technical and engineering stages of the innovation process are made systematic through experimental development. When technological change is sustained by systematic R & D, it is more likely to become the source of sustained economic growth than when it comes about haphazardly from trial-and-error and 'learning-by-doing' in the production process.

The concept of science policy originated from the need to direct scientific research toward specific social objectives rather than the pursuit of science for its own sake, and to make the application of research results to concrete technological problems more efficient. During World War II, science policy focused on the development of new military technologies. After the war, science policy was also concerned with the development of new technologies of production, at first to accelerate the reconstruction of war-damaged economies, but later, in the face of increasing economic competition, manifested in the so-called 'technology gap' between different national economies. It was at this time, too, that the idea arose of consciously directing the application of science to the specific economic problems of the developing countries where the inadequacy of local scientific capacities was seen as an obstacle to innovation and economic growth sustained by technological change.

'Science' is used here to refer to a body of knowledge and methods of observation and measurement, experiment and deduction which are used to create new theories. Acquisition of this knowledge and the exercise of these methods takes place in specialized institutions, thus enabling learning and the gathering of new facts to occur more rapidly. Certain other specialized activities are required to support these basic scientific functions; there are more or less conventionally accepted definitions, described below, of these activities.

The relevance of scientific knowledge and education and the application of scientific research methods to society's technological needs depend on the relationship between the specialized scientific institutions and the production system. Our discussion in the preceding section on 'Technology Policy' referred to some of the conditions which may inhibit the modern industrial sector in a developing country from initiating technological innovation based on local R & D. In China, where pre-modern science and technology were both highly advanced, it is particularly interesting to speculate on why science did not become the basis of technological innovation. Some of the literature listed in Appendix 2 on traditional science and technology considers this question, both from the angle of traditional social and economic structures in the production system, and with respect to the nature of the science itself.

There has been a tendency in both developed and developing countries to claim for the institutional forms of science the universal validity attributed to scientific knowledge and method, not recognising that these institutional forms are the specific product of the societies in which they developed—that is, post-Renaissance Europe and the European-derived cultures of North America. When they are transferred to quite different social and cultural contexts in the developing countries, institutions on this model may establish new linkages with institutions in the production and the political systems of the new environment only slowly, not at all, or on a basis quite different from that of the relation between science and society in developed countries. Because the institution of modern science and the institutions of modern industrial production are usually transferred separately to the developing countries, institutional links between the generators of new scientific ideas and the industrial application of these ideas may not be developed. The 'science system'—the community of scientists and the institutions in which they perform scientific activities—may then become an isolated enclave within the developing country.

The concept of science as a purely academic pursuit, existing for the discovery of new knowledge—a concept often sustained by established social patterns—also reinforces the institutional separation of research from production. The orientation of the scientific community, its linkages and its patterns of external communication may all be directed toward the scientific profession in other countries, usually the developed countries, where advanced science and 'frontier' research are centred. The ultimate expression of this orientation is 'brain drain', that is, the migration of scientists from developing to developed countries. Science policy makers in a developing country, therefore, may be faced with either little or no indigenous science, or persistently low quality of scientific work, or with too much science and a flourishing scientific community whose work has little bearing on the needs of local economic and social development.

A *Policy for Science*

The science policy of any country has two essential objectives: to foster and maintain a viable and creative scientific community, and to direct scientific research toward the development of new technologies for economic, social and military purposes. Policy *for* science is expressed in the allocation of resources for the performance of scientific activities, including the support of basic or non-oriented research, scientific education and training, applied research and experimental development, and specialized R & D-supporting and R & D-related activities. Especially when total resources for science are limited, competing claims for finance, scientific manpower, laboratory equipment and library services cannot all be satisfied, and in the resolution of these claims and the allocation of resources by the policy-makers, priority may have to be given—implicitly if not explicitly—either to scientific advance or to the application of existing knowledge to concrete economic and social problems.

The development context tacitly emphasizes the second objective of science policy: the application of science to technological advance. Even so, some resources must be devoted to building scientific capabilities for the future: some basic research is essential not only because the results of such research may have future technological application, but also to maintain the vitality of the scientific community, whose inherent purpose is the discovery of new knowledge through research.

The national government is the effective science policy-maker in most countries, even in the absence of a formal policy for science or an official ministry or department of science; this is because the national government is usually the major source of funds for scientific activities. The allocation of government resources among different research programmes, to different fields of science and to different research institutes involves not only a choice between scientific development *per se* and science as the instrument of technological policies, but also the policy-maker's perceptions of the kinds of institutions, the patterns of organization, and the definition of research programmes which will facilitate their objectives.

There has been comparatively little data forthcoming from official Chinese sources on these issues. The fragmentary information available has been assembled in a number of studies to construct a picture of the pre-Cultural Revolution organization of science policy-making institutions and some of the institutions formally responsible for performing scientific activities. Although estimates have been made of science budgets, manpower, etc., the picture which emerges from these studies does not yet make clear the dynamics of the science system and of science policy making. Of these studies J. M. H. Lindbeck, 'The Organization and Development of Science' (entry no. 154), Surveys and Research Corporation, *Directory of Selected Scientific Institutions in Mainland China* (entry no. 196), and Yuan-li Wu and Robert Sheeks, *The Organization and Support of Scientific Research and Development in Mainland China* (entry no. 210) are the basic reference works. Most of the other entries in this category either precede these studies and have been superseded by them; are derivatives from these studies, drawing upon the information compiled in them; or are of more limited scope. Of studies of science policy and science institutions in China since the Cultural Revolution, the most comprehensive are John A. Berberet, 'Science and Technology in China' (entry no. 118) and Leo A. Orleans, 'China's Science and Technology: Continuity and Innovation' (entry no. 184).

## B *Policy Toward Science*

Sociologists of science have long recognized the importance of institutional structures, the role of the scientific community versus other groups in society, and the effectiveness of various systems of rewards and incentives as factors affecting the performance of research and other scientific activities. In all countries, and perhaps especially in developing countries where science is an alien transplant from another culture, the scientific community may be regarded with a certain ambiguity by other parts of the society. Where the success of economic and social policies requires co-operation from the scientists in developing new technologies, policy-makers may manipulate the social and political conditions in which the scientists work to ensure this co-operation.

Government control of the bulk of financial resources for science also constitutes a source of control over the kinds of research to be performed and the activities of the professional scientific community, although this may be tempered by the belief that a certain degree of academic freedom is necessary to maintain scientific creativity. It follows that the more important science is to the attainment of the government's economic, social, military and political objectives, the greater may be the influence of the scientific community on the policy-makers, at least where its own interests are concerned. Alternatively, in such circumstances the government may exert extraordinary political and/or economic pressures on scientists to enforce its own science policy and other objectives. Thus, a not insubstantial component of science policy consists of a policy *toward* science: the relationship between the planners and policy-makers as a political group, and the scientific community.

The political and social position of the scientific community in the People's Republic of China has received considerable attention in the literature, predominantly centring around the effort to make scientists both 'red' and 'expert'. With few exceptions, these studies treat Chinese policy toward the scientists, as opposed to policy for science, in terms of its political objectives rather than in terms of development and technological advance. For example, political structures in China have enabled the authorities to control communications between the Chinese scientific community and its professional counterparts in other countries; fluctuations in these communications have been interpreted as changes in political policies toward science. For references to this aspect of Chinese science policy, see the entries in Section IV on 'Scientific Communication'.

The political and sociological aspects of Chinese science policy have been most extensively treated in R. P. Suttmeier, 'Party Views of Science: The Record from the First Decade' (entry no. 254), and the same author's doctoral dissertation, 'The Chinese Academy of Sciences: Institutional Change', Indiana University, 1969.

## A Policy for Science

115 ARTS AND SCIENCES IN CHINA
*Science Notes*. 1(1), 38-9, 1963; 1(2), 39-40, 1963; 1(3), 39-40, 1963; 1(4), 26-7, 40, 1963; 2(1), 40, 1964.
Brief review of current scientific developments in China.

116 BARNETT, James W. Jr.
What price China's bomb? *Military Review*, 16-23, Aug 1967.
Estimate of costs of Chinese nuclear development program, compared with costs of purchasing chemical fertilizer plant, increasing steel capacity, developing energy resources (electricity, petroleum), developing a paper industry, or improving transportation. Capital investment and operating expenses of major operations in the creation of a nuclear device, calculated in terms of corresponding U.S. operations: 'The Chinese nuclear program from 1957 through the first explosion in 1964 cost an estimated 2.5 billion U.S. dollars. The current annual cost is 470 million U.S. dollars . . . .' 'By converting estimated outlays for nuclear development into productive capacity equivalent, one can better understand what the Chinese have denied themselves in future production of consumer and producer goods by electing to expend their resources for an economically non-productive program.'

117 BERBERET, John, A.
The prospects for Chinese science and technology. *TEMPO* 68TMP-26. Santa Barbara, General Electric Company, Feb 1968.
'In this discussion the prospects for science and technology in Communist China will be examined briefly in a time context. The foundations that have been and are being laid for progress there today will be emphasized much more than the progress that has been made.' Section 3--China's scientific and technological heritage. 4--An overview of contemporary Chinese science and technology. 5--The training of scientists and engineers; the political treatment of Chinese intellectuals; significant scientific and technological developments; major inadequacies in Chinese science and technology. 5--Effects of the Great Proletarian Cultural Revolution. 6--Prospects for scientific and technological advancement.

118 BERBERET, John A.
Science and technology in China. *Current Scene*, 10(9), 12-19, 1972.
Organizational lines; professional manpower; Soviet assistance; research and development. Conclusions suggested by reports from foreign scientists.

119 BERBERET, John A.
Science and technology in Communist China. *TEMPO Research Memorandum* RM 60TMP-72. Santa Barbara, General Electric Company, 8 Dec 1960.
'It is the purpose of this paper to make additional analyses of the development of Chinese (professional) manpower' . . . the significance of this manpower development to the scientific and

technological capabilities and subsequent military development of China. 'From these manpower studies and studies of the activities and accomplishments of the Chinese scientists and engineers to date, some conclusions about the capabilities of the Chinese to develop nuclear armaments are made.' Section I--Contemporary China. II--The Chinese educational system. III--Science and technology: the administration of scientific and technological programs; the China Academy of Sciences; the Research Institutes of the Ministries; research and engineering in the universities and colleges; professional societies; scientific and technical journals; Russian assistance; research and development in selected fields. IV--Scientific and technical manpower: summary. V--The Chinese People's Republic and nuclear armaments. VI--The future and China.

120 BERBERET, John A.
Science, technology and Peking's planning problems. *TEMPO* RM 62TMP-91. Santa Barbara, General Electric Company, Dec 1962.
Section 2--Economic and technological shortcomings: agriculture; industry; capital for foreign exchange. Section 3--Science, technology and the government: the government regards science and technology as basic to the development of an industrial society but subjects scientists and engineers to political pressure. Section 4--Current status of science and technology: Chinese have knowledge of very recent technologies, but low capability to use advanced technology. Section 5--The development of nuclear weapons. Section 6--Factors affecting the future of science and technology: greater emphasis on applied science, no shortage of science and engineering manpower at present level of the economy; not likely to produce notable accomplishments in basic science.

121 BERNAL, J.D.
A scientist in China, Part I. Universities and colleges. Part II. *Nation,* 49(1255), 424-6, 1955; 49(1256), 463-4, 1955. Also Science in China. *United Asia,* 8(2), 129-33, 1956.
Observations based on a 1955 (?) visit to China at the invitation of Academia Sinica. Visits to 60 Chinese institutions; universities and research laboratories, factories and works. 'My general impression was that, here in China, . . . the grafting of the new scientific culture . . . had already been achieved smoothly and in an incredibly short time.' 'In the natural sciences and in technology the immediate tasks are those of assimilating existing knowledge and technique and using it in the building up of the economy. Teaching in educational institutions is "essentially an adaptation of the Soviet system"; professors mostly had British or American training. Research developed more rapidly than teaching research in Academia Sinica; industrial research, survey of natural resources, agricultural research.

122 BOORMAN, Howard L.
The scientific revolution in Communist China. *China Report,* 4(5), 10-20, 1968.
Problem of establishing a baseline from which to measure development of science in China. Chinese Communists' goal of industrialization and modernization require replacement of traditional ideas,

beliefs and myths about nature with scientific concepts and introduction of modern, scientifically-based technology. Chinese commitment to science at two levels: development of advanced science and popularization of science.

123 BRULE, Jean-Pierre
*China comes of age* (La Chine a vingt ans). Rosemary Sheed (trans.). Harmondsworth, Penguin Books, 1971.
General discussion of population growth, economic development and military strategy in China since 1949. Chap. 5--The nuclear challenge. Review of Chinese science policy. 'The reason for China's scientific policy is her own determination to create (nuclear weapons).' Return of Chinese scientists from the West after 1949. The science policy-making functions of the CAS supplanted by the Party in 1958; political concessions made to scientists. The 12-Year Plan (1956-67) and the 10-Year Plan (1963-72) for scientific development. Higher education. R&D is 1.1% of GNP. Import of Soviet scientific aid in connection with nuclear development; imports of scientific hardware from Japan and Europe; purchase of scientific and technical literature. Advances in certain fields of science. 'By considering the development of people's China as they would that of a country with a modern economic infrastructure, none of the (Europeans) saw, or indeed wished to see, what was actually happening, scientifically and industrially, in China. Given her position in the world, China was determined to sacrifice absolutely everything to the gods of nuclear energy.' Summary of nuclear missile development programs. The potential political impact of China's emergence as a nuclear power.

124 *BUSINESS WEEK*
China's push to catch up in science. No. 1901, 116-20, Feb 1966.
Survey of Chinese science based on interview with Cheng Chu-yuan. China's self-imposed isolation has a 'devastating' effect on attempts to speed development of science and technology; expansion of numbers of scientists, engineers and facilities at the sacrifice of quality. 'Chou En-lai controls science' . . . through the State Scientific and Technological Commission and the CAS. New research branches established in nuclear physics, semiconductors, electronics, automation, high-polymer chemistry; radio-electronics; ultrasonic technology. Technological advances include chemical, pharmaceutical, petroleum, iron and steel and heavy manufacturing industry. Military developments.

125 *BUSINESS WEEK*
Red China's beehive of science. No. 1541, 183-4, 186, 190, 192, Mar 1959.
General observations on Chinese scientific research, science policy, education, based on the observations of Dr. J. Tuzo Wilson. 'Scientific observers . . . have been surprised by the weight given to basic rather than applied research.' The institutes of the CAS and university research researchers concentrate on basic research. 'What applied work there is falls largely on the government's scientific ministries, but these agencies employ far fewer scientists than the Academia. Results

of the fundamental work are not always up to Western standards, but there is evidence they soon will be.' Organization of geology. Lab equipment. Higher and secondary education. Scientific libraries, journals, publication (Chinese publish 11% of scientific papers in Iron Curtain countries).

126 *BUSINESS WEEK*
Red China's 'leap' toward science. No. 1734, 102-4, 106, 108, Nov 1962.
China has made some progress in developing its scientific capabilities, but is still backward compared to the modern West. 'Apart from the twin problems of too few trained men in too few adequate labs, Chinese science has been hurt by the demands of the Korean War and by drastic domestic programs of change; it has suffered from the Communist Party's tight supervision of all intellectuals. The biggest handicap of all has been the government's demands for impossibly fast research results. Most of Red China's scientists and research institutes are concentrating, under order, on industrial rather than theoretical research problems.' The Academy of Sciences; CAS budget; R & D budget; Scientific Planning Committee; Scientific and Technological Commission. Sketches of Ch'ien Hsueh-shen; Wang Kan-chang; Ko Ting-sui; Ch'ien San-ch'iang. 'Western experts . . . agree that . . . few important developments are likely to emerge from Chinese laboratories until well into the 1970s.'

127 *CHEMICAL ENGINEERING PROGRESS*
Nuclear progress in Red China. 56(3), 16-17, 1960.
Manufacture of scientific equipment, instruments in China. Sino-Soviet cooperation: the Sino-Soviet Agreement on Scientific and Technical Cooperation (1954); joint study of 122 scientific and technical problems. Scientific and technical cooperation with Eastern Europe, e.g., at Dubna. China's 12-Year Plan for science and technology.

128 CHESNEAUX, Jean
The spread of modern science in the Far East. *A general history of the sciences*, Rene Taton (ed.), Pt. VI, Chap. 8. London, Thames and Hudson, 1966.

P. 597-599: scientific advances in modern China (1911-1949); science in the People's Republic of China.

129 CH'IEN, Cheng
A dragon with nuclear teeth. *Military Review*, 78-84, Oct 1966. Commentary on Chinese science policy and military applications of science by a colonel in the Republic of China Army. Soviet collaboration in the 12-Year Plan for science and technology abrogated in 1960 with Soviet withdrawal of scientists and technicians. Chinese policy thereafter shifted from establishment of research institutes and facilities to manpower: development of research in colleges and universities; enlistment of scientists and technicians from abroad; reappraisal of scientists earlier associated with the Republic of China. The State Scientific and Technical Committee; the CAS; the Academy of Military

Science under the Ministry of National Defense; the Chinese University of Science and Technology. The nuclear research program; rocket development.

130 *CHINA NEWS ANALYSIS*
Sciences in China. No. 131, 11 May 1956.
The Academy of Science: history, development, learned committees, north-east, north-west, the next 12 years, training of scientists, Russian relations, Marxist natural sciences.

131 *CHINA NEWS ANALYSIS*
Scientific research. No. 132, 18 May 1956.
The ghost at the scientific feast. Physics, mathematics, chemistry, biology, physical geography, technical sciences, philosophy, social sciences, scientific journals.

132 *CHINA NEWS ANALYSIS*
Scientific work. No. 263, 6 Feb 1959.
A realistic policy aimed at raising the technical level in rural areas and small towns, while maintaining the major modern scientific institutes, and at cutting expenditure for scientific work. Country science: local and provincial scientific research organs; the new scientists and the mass line. The scientific world: reform of scientists; organization of science; the party and science. Scientific equipment.

133 *CHINA NEWS ANALYSIS*
Tools for research. No. 504, 12 Feb 1964.
Material conditions in China do not facilitate pioneering research; thus far China is only in remote contact with the world of science and modern techniques and gives no sign of any technical leap forward. Rewards for inventions. Five Year Plans: resources allotted for purchase of scientific instruments; production of instruments; repair; exchange of equipment; laboratories; professors. Appendix: science literature.

134 *CHINA NEWS ANALYSIS*
University professors. No. 97, 26 Aug 1955.
Scientific research, p. 6-7.

135 CHRISTIANSEN, W.N.
Science and the scientist in China today. *Eastern Horizon*, 7(2), 36-40, 1968.
Revised version of article originally appearing in *The Australian Physicist*. Brief review of history of Chinese science and technology; Russian influence on Chinese higher education and scientific research. Effect of Russian withdrawal. Science subordinate to political goals. Reform of scientists.

136 *CURRENT SCENE*
Chinese science on the mend. 9(8), 17-18, 1971.
Reappearance of Chinese officials in charge of scientific organizations after the Cultural Revolution; list of foreign scientists visiting China. Shift in focus from theoretical to applied research.

137 DEAN, Genevieve C.
Science and the thought of Chairman Mao. *New Scientist*, 45(688), 298-9, 1970.
The Maoist strategy for applying science to the problems of modernization calls for reorganization of the scientific establishment, general education in science.

138 DELEYNE, Jan
*L'Economie Chinoise*. Paris, Editions du Seuil, 1971.
IV--L'agriculture: mechanization; chemicalization; research in agronomy; popularization of modern methods. V--La production industrielle et les transports: petroleum, nuclear and armaments, iron and steel, electronics, machine tools and chemical industries. VI--Les progres techniques et scientifiques: brief review of science and technology and history of science in China; role of scientific research in economic development; organization of research (CAS); research in universities and industrial ministries (354 research institutes in 1963); coordination of industrial research in factories and universities, the system of secondment of personnel; research manpower; research and politics. Industrial technology: summary of impressions reported by foreign visitors; self-reliance and rejection of foreign technical assistance; short-term presence of European and Japanese businessmen and engineers in China; foreign industrial and technical echibitions in China; systematic exploitation of foreign (esp. North American) publications; isolation of Chinese science and technology; foreign technology copied; recent technological achievements. Details on Peking No. 1 Plastics Factory (1966). Manufacture of coke and chemical products in Peking (1966).

139 EITNER, Hans-Jurgen
Erziehung und wissenschaft in der Volksrepublik China, 1949 bis 1963. *Gesprachkreis Wissenschaft und Wirtschaft* BDI/DIHT/SV, 1964.
Education and science in the Chinese People's Republic, 1949-1963. I--From Confucianism to Communism. II--System of universal education. III--College system. IV--Scientific academies: Chinese Academy of Sciences, medical academy, Chinese Academy of Agricultural Sciences. V--Research institutes. VI--Scientific societies for natural sciences and technology. VII--Leading libraries (outside the CAS and colleges). VIII--On the general situation of science and technology. IX--Financing of education and science.

140 FEINBERG, Betty
Science in China: report on the AAAS symposium. *The China Quarterly*, No. 6, 91-7, 1961.

141 FISHANE, S.J.
China's technical position. *Contemporary Review*, Vol. 197, 154-5, 1960.
Brief comment on the scientific and technical capability of China. 'All in all, it seems that Communist China has a long way to go in science and industry.' Lack of trained scientists in China and their disaffection; 'the expansion of science in China seems largely dependent upon Russia.' Soviet Union has not given China full information for nuclear research.

142 FOO-KUNE, C.F.
*Science and industry in China.* Address to ASLIB Technical Translation Group meeting, 9 Dec 1960. Unpublished typescript.
Brief review of the scientific establishment in China from 1949 through the Great Leap Forward.

143 GOULD, Sidney H.
*Sciences in Communist China.* Washington, American Association for the Advancement of Science, No. 68, 1961.
Papers prepared for a symposium held in New York, 26-27 December 1960.

144 HAMBRAEUS, Gunnar
Science and technology in China after the Cultural Revolution. Ingenjörsvetenskapsakademien. *Tidskrift för Teknisk-Vetenskaplig Forskning,* 43(4), 149-54, 1972.
Visit to China by Swedish scientist's, 25 Feb-11 Mar 1972. Reorganization of science as a consequence of the Cultural Revolution: Academia Sinica directed by a revolutionary committee. 2/3 of 120 CAS research institutes are now under provincial governments or dual Academia/local authority, although still financed by the Academia and co-ordinated with Academia's policy. Priority assigned to research on basis of how promising an idea looks for development and its industrial and agricultural applications. 'Basic science' . . . is an important function of the Academia. Publication by the Academia. Communication with scientists in other countries, but 'science cannot be separated from politics.' 'Re-education' of scientists in May 7th schools. Scientific instruments. Research in physics.

145 HARARI, Roland
The long march of Chinese science. *Science Journal,* 4(4), 78-84, 1968.
Article based on interview with four French scientists recently returned from China. China is advanced in some fields of research but remains a scientifically underdeveloped country. The Chinese government respects scientific thought and method but requires that science and technology must result in practical economic application. New fields of science may develop in China, due to the isolation of Chinese science and to the different problems being tackled by science in China.

146 HASKINS, Caryl P.
*The scientific revolution and world politics.* New York, Harper & Row, for the Council on Foreign Relations, 1964. Chap. 6--Science and the liberal frame: the Soviet Union and China, p. 81-105. Also Where is science taking us? *Saturday Review,* p. 58-60, 6 Jun 1964. Also Science in Red China: the enigmatic dragon. *Air Force and Space Digest,* p. 60-3, Aug 1964.
Technological achievements in contemporary China are natural extensions of China's history of technological development; they are 'more expressive of energy and will than of striking originality,' adopted from Western technology through the Soviet Union. Scientific achievements in some areas are 'remarkable' and an index c

originality and key to the longer future. Manpower shortage; training in the Soviet Union and in Chinese institutions of higher education. The gulf between applied and pure research. Science policy planning and institutions: the State Council's Planning Committee for Scientific Development; the CAS; the Twelve Year Science Plan. 'The importance of Soviet philosophy and Soviet help . . . in the shaping of Chinese science is one of the most conspicuous features of the whole pattern.' 'It was absolutely inevitable that the Communist Chinese should turn to the Soviets both for material help and, more lastingly, for philosophy and methods of organization of a state science.' The Chinese regarded Soviet assistance as an initial send-off, aimed at achieving technical and scientific independence. Chinese science and technology 'may suffer a considerable immediate setback' as a result of the Sino-Soviet rift, but are 'likely to emerge, ultimately, as yet more powerful elements in a nation scientifically, as politically, an astute student of Russian methods and patterns, but operationally independent of them.' China can still profitably use much existing science and technology, but whether original science can flourish under totalitarianism remains to be seen.

147 HINSHELWOOD, Sir Cyril
A visit to China. *New Scientist*, 6(156), 858-60, 1959.
Report of a visit to China in 1959 by the President of the Royal Society. Organization of science: the Academy of Sciences. Higher education: without neglecting basic sciences, practical application is emphasized at an early stage. Description of research programs.

148 *INDUSTRIAL RESEARCH*
Red China science focuses on short-term. 13(7), 23-4, 1971.
Brief article reporting observations of two American scientists (Galston and Signer) visiting in China in 1971. Use of gibberellin in fungus form to increase crop yields. Use of acupuncture as anaesthetic. Research in herbal medicines. 'Their report . . . showed that, despite its isolation from the West, Chinese science is making impressive strides.' The Chinese 'believed in pure research aimed at the long term . . . but felt that the short-term needs were much more pressing; in the future, they said, they are reconciled to the necessity of imitating or purchasing scientific research in the hope of catching up.'

149 ISBERG, Pelle
Vetenskapen i Kina. *Teknisk-Vetenskaplig Forskning*, 32(2), 53-7, 1961.
Natural sciences in China. Based on AAAS conference on Chinese science: mathematics, physics, chemistry, geology, mineralogy, meteorology, engineering sciences, astronomy, botany, zoology, genetics, physiology (medicine and pharmacology), biological sciences.

150 JAUBERT, Alain
Recherche et developpement en China. *La Recherche,* 2(11), 339-49, 1971.
Based on available published sources and the testimony of 'observers whose positions do not permit publication of their names.' Reform of the education system during the Cultural Revolution. Reform of scientific attitudes and method, from pure research to applied research and invention by workers. Self-criticism of scientists. The Second Revolution in Production: technological development and innovation in the Cultural Revolution. Medicine and public health. Achievements in scientific research and industrial developments in the 1960s. The Cultural Revolution and agriculture.

151 KOBAYASHI, Fumio
Organization of scientific and technical research. *Long-term prospects for the Chinese economy,* Pt. III, Chap. IV, Shigeru Ishikawa (ed.). Tokyo, Asian Economics Research Institute, No. 119, 1967.

152 KUSANO, Fumiko
Evaluation of comprehensive study of Chinese Communist science and technology. Tokyo, 27 Sep 1964. Trans. *JPRS,* 37, 828, 26 Sep 1966.
1--Science and technology and the basic policy on education. II--Education of specialists and the state of scientific research during the period of the first Five Year Plan. III--The plan for scientific and technological progress and the mechanics of research and dissemination. IV--The second Five Year Plan and the effect of reliance on Soviet science and technology. V--The anti-science and technology policy during the period of economic adjustment. VI--Policy for the development of science and technology for self-reliance. VII--Critical evaluation of science and technology in Communist China.

153 LINDBECK, John M.H.
An isolationist science policy. *Bulletin of the atomic scientists,* 25(2), 66-72, 1969.
Chinese interactions with the external world since 1950 have gone through several stages: massive borrowing from the Soviet Union, 1950-60; the break with the USSR followed by limited and exploratory policies of finding non-Communist substitute sources for Russian scientific and technical assistance, 1960-1966; and the breakdown of scholarly and scientific communications in the Cultural Revolution.

154 LINDBECK, John M.H.
The organization and development of science. *Sciences in Communist China,* Gould (ed.). AAAS, No. 68, 3-58, 1961. Also *The China Quarterly,* No. 6, 98-132, 1961. Also *China under Mao: politics takes command,* Roderick MacFarquhar (ed.), p. 333-67. Cambridge, M.I.T. Press, 1966.
Conclusions: China possesses a retarded scientific culture; emphasis on applied sciences, engineering, and technology in support of national development programs; basic research is conducted for pragmatic motives; China will not soon be a major

contributor to theoretical knowledge; support given to scientific research is still modest by international standards; Soviet assistance and guidance have given a Soviet imprint to Chinese science; development of large and top-heavy administrative superstructures; limited political influence of top scientists; the presence and supervisory intervention of the Communist Party is evident throughout the entire range of scientific organizations; the social sciences are neglected. Priorities in developing science (1960): research related to development of the national economy; research on pioneering science and technology; research on fundamental theories. Appendix: regulations on the organization of the departments of the Chinese Academy of Sciences. Tables: budgeted expenditures for science; growth of scientific research institutes and personnel (1949-1959); organization of science in Communist China; organization of the Academy of Sciences; institutes and research centres of the CAS, by department; organizational structure of research institutes.

155 LINDBECK, John M.H.
Chinese science: it's not a paper atom. *New York Times Magazine*, 38-9, 60, 62, 64, 67, 70, 8 Jan 1967.
Chinese science policy attaches highest priority to nuclear weapons development, missile development. By 1959, China had a corps of skilled scientists and engineers, mostly trained abroad, the necessary capital and industrial resources, and the requisite raw materials and special technological facilities for the development of nuclear weapons. Organization of military science: the Academy of Military Sciences; the Chinese People's Liberation Army's Scientific and Technological Committee of National Defense. Costs to industrialization on concentration on military science; benefits to development of science stemming from its military utility. Industrial technology in China copies older technologies and methods imported from abroad; no comprehensive programs of advanced industrial research and development.

156 MACIOTI, Manfredo
Developpements recents de la science et de la technologie en Chine Communiste. *Etudes et analyses*, No. 89. Brussels, C.C.E., 15 Jan 1970.
Introduction: general, political, social, economic aspects. Science and technology: historical framework, basic statistics, atomic sector, missile sector, aeronautics, electronics, electric power, transport, chemistry.

157 MACIOTI, Manfredo
Science et technologie de la Chine Communiste. *Etudes et analyses*, No. 57. Brussels, C.C.E., 30 May 1969.
Introduction: politics, society, economy, education. Science and technology: nuclear arms, electronics, machine industry. Technological forecast.

158 MACIOTI, Manfredo
China uses science policy 'to walk on two legs'. *Science Policy News*, 2(6), 70-1, 1971.
Condensation of recent articles on science policy and education in China.

159 MACIOTI, Manfredo
Hands of the Chinese. *New Scientist,* 50(755), 636-9, 1971.
'It is the purpose of chis article to attempt to assess the scientific and technological capability of the People's Republic of China today. To this end, I shall try and define the position of China in relation to the rest of the world in the broad areas of culture, education and science, as well as in four advanced technologies (nuclear, missiles, jet aircraft and computers). The main conclusion of this survey is that China is emerging as the third scientific and technological power of the world, although this rank might be challenged by Japan.'

160 MACIOTI, Manfredo
Gli scienziati a piedi nudi. *Successo,* 115-18, Jan 1971.
Scientists go barefoot. 'The aim of this article is to attempt to assess the scientific and technological capacity of Communist China today. For this purpose we will take two general indicators: the level of higher education and the investments made for scientific research, in comparison with the situation in the rest of the world. We shall therefore see what is the status of four fundamental technologies (nuclear, missiles, jet planes, and computers) in comparison with international levels.' '. . . China's scientific progress indicates that the country's limited financial and human resources have been concentrated on the principal objectives that were set forth in the national programs.' Mao's policy aims at modernizing the country by starting at the roots; has a propensity for applied sciences and technology.

161 MACIOTI, Manfredo
Il sistema della ricerca in Cina. *La Critica Sociologica,* Vol. 22, 213-17, Summer 1972.
Brief review of scientific achievements in China. Lists of primary and secondary sources of information.

162 MACKAY, Alan
Science in Asia. *Asia, a handbook,* Guy Wint (ed.). London, Anthony Blond, 1965. Revised as *Asia Handbook,* Harmondsworth, Penguin, 1969.
P. 626-628 (640-642): China. 'Science comprises two basic activities. The first is science as an intellectual activity . . . the second of the activities of science is the application of known principles in new situations.' Scientific manpower in China. The Academy of Sciences is responsible for basic science and pure science. The Twelve Year Plan for science. 'The Chinese system has now 'taken off' and is self-sustaining.' 'Science will be an agent of this transformation (of China into a modern industrial nation of 1,000 million).'

163 MARU, Rushikesh
*Research and development in India and China: a comparative analysis of research statistics and research effort.* Project on the comparative study of research and development in India and China, Occasional Paper No. 1 (preliminary draft). Delhi, Research Policy Program, Lund and Centre for the Study of Developing Societies, unpublished typescript, 10 Apr 1969.

The framework for comparing R & D policies and performance in India and China is their common problems and different social systems. The first step is to compare their efforts to develop R & D potential--by measuring input in terms of national expenditure on and manpower engaged in R & D. Non-availability of statistical information limits comparison to only a few dimensions of R & D expenditure and manpower.

164 MENDELSSOHN, Kurt
Science and technology in China. *Eastern Horizon*, 5(7), 5-11, 1966.
Science in the Western sense hardly existed in China before 1950. Science education in universities; technical training in technical institutes and specialized institutes; the educative value of backyard workshops overrides low productivity. Shortage of scientific manpower is a greater problem than shortage of equipment. Owing to official encouragement, general enthusiasm and high average intelligence, the climate in present-day China is very favourable for a rapid scientific and technological expansion.

165 MENDELSSOHN, Kurt
Science in China. *Nature*, 215(5096), 10-12, 1967.
Organization of scientific and technical education and research. The Russian influence on education and research, with some important changes in the Chinese case. Description of modern technology observed by author on trips to China in 1960 and 1966. Effects of changes in the wake of the Cultural Revolution cannot yet be measured.

166 MODELSKI, J.A.
Communist China's challenge in technology. *Australian Quarterly*, 30(2), 57-68, 1958.
General survey of Chinese science policy: application of science to military and industrial goals; trained and educated manpower is greatest potential for technological change. Expansion of state expenditure on research, increased enrolment in higher education. Priority on development of nuclear energy and rockets. 'Communist China has thus recently embarked on a huge scientific and educational expansion programme, calculated to give her the status of a leading technological power within a decade or two, and full nuclear potential by the middle 1960s.'

167 NEEDHAM, Joseph
Chinese science revisited. *Nature*, 71(4345), 237-9, 1953; (4346), 283-5, 1953.
Pt. 1--Observations based on the author's 1952 visit to China. Descriptions of the organization of the Chinese National Academy (Academia Sinica), universities, research in the Ministry of Health, National Institute of Biological Products; progress in public health. Pt. 2--Research programs emphasize practical problems. Attention to education and popularization. Review of research programs. 'In accordance with the world-outlook which has triumphed in China, Chinese scientists are greatly interested in dialectical materialism. The fact that there is very little opposition to it is noteworthy--the conceptions of Chinese humanism are fusing with dialectical materialism . . .'

168 *NEWSWEEK*
Wither Red China's 'march on science'? Vol. 57, P. 76, Jan 1961.
Summary of papers read at AAAS symposium on Chinese science: on manpower; nuclear energy; natural resources; public health; electronics; heavy industry; meteorology; medicine.

169 *NEW SCIENTIST*
A look at China's economic potential. Notes on the news, 40(617), 6, 1968.
Brief review of findings by Yuan-li Wu that 5% of Chinese GNP in 1964 was spent on science.

170 NIELSEN, Robert B.
*Scientific, academic, and technical research organizations of Mainland China; a selective listing.* Library of Congress, Aerospace Technology Division, 1965, rev. Sep 1966.
'The purpose of this listing has been to provide a guide to standardizing the translated names of organizations . . . . The listing is not considered comprehensive.' Scientific, technical, industrial, and academic organizations of Mainland China. Institutes and facilities of the Chinese Academy of Medical Sciences. Institutes and facilities of the Chinese Academy of Sciences. Institutes and facilities of the Chinese Academy of Agricultural Sciences.

171 NIIJIMA, Junryo
*China's Cultural Revolution and technical reforms.* Chugoko Shiryo Geppo (China Research Institute), No. 127, 25 Nov 1958. Trans. *JPRS,* 673-D, 23 Apr 1959.
I--Viewpoint. 'Cultural revolution' and 'technical reform', as used by Liu Shao-ch'i, May 1958 Report to 8th NPC, 2nd session. II--The Scientific and Technical Research System: the mechanism for guidance of research; the organization of scientific and technical research; research funds; research staffs; scientific and technical interchange and the collection of scientific information. III--The Cultural Revolution. IV--China's technical level: the scientific and technological level of Old China; power technology; extractive technology; materials technology; machine technology; construction technology; communications technology; traffic technology; controls technology; cultivation technology; conservation technology; health technology. Technology of the new China, divided into the 12 categories of Yoshiro Hoshino and measured by a) whether China has achieved world's latest technology; and b) by the relationships among fields. V--The Mass Technological Revolution (especially agriculture). VI--The significance of the Cultural and Technological Revolution. Details and statistics compiled from original Chinese sources.

172 *NOTES ET ETUDES DOCUMENTAIRES*
Organisation et developpement de la science en Republique Populaire de Chine. Paris, Secretariat General du Governement, Direction de la Documentation, No. 3255, 18 Jan 1966.
Initial resources; control of science; the Academy of Sciences; planning of research; development of the means for research;

formation of trained manpower; science 'for the masses'; a few results which have been attained; science and national defense.

173 OLDHAM, C.H. Geoffrey
*A.A.A.S. Symposium on Chinese Science.* Pt. II: China's Scientific Revolution. Institute of Current World Affairs, unpublished newsletter CHGO-2, 8 Feb 1961.
Review of papers on geology, mining and metallurgy, electronics and computing, astronomy, botany, zoology, genetics, geophysics and physics given at A.A.A.S. Symposium. Conclusion: China is in the midst of a scientific revolution; at the moment is relatively backward but will rapidly become one of the most scientifically advanced nations in the world.

174 OLDHAM, C.H. Geoffrey
China today: science. Cantor lecture to the Royal Society of Arts, 25 Mar 1968. *The Journal of the Royal Society of Arts,* 116(5144), 666-82, 1968.
Growth of science in China to 1966; science linked to objectives of the Communist government; indoctrination of peasants and workers with scientific knowledge. General assessment of Chinese science. Science in the Cultural Revolution; the Cultural Revolution and military research; political conflict over science, technology and economic development. Schools and universities closed down in the Cultural Revolution. The priorities have changed from 'catching up' with advanced countries' science to concentrating scientific effort on problems of direct relevance to China's development needs.

174a LOW, Ian
Where Chinese science is going. *New Scientist,* p. 31-2, 4 Apr 1968.
Based on Cantor lecture to Royal Society of Arts by C.H.G. Oldham, 25 March 1968.

174b *NATURE*
Science after the Cultural Revolution. 217(5135), 1196-97, 30 Mar 1968.
Review of the Cantor lecture to the Royal Society of Arts, 25 Mar 1968, by C.H.G. Oldham.

174c SULLIVAN, Walter
A 'science revolution,' too, under way in China. *New York Times,* 21 Apr 1968.
Article based on Cantor lecture at Royal Society of Arts by C.H.G. Oldham.

175 OLDHAM, C.H. Geoffrey
Chinese science and the Cultural Revolution. *Technology Review,* 71(1), 22-9, 1968.
International contacts with foreign science flourished 1964-66, but were broken during the Cultural Revolution. History of Chinese science and technology. Political attacks on leaders of the Academy of Sciences. Policy conflict on choice of technology, importation of foreign technology, defense research and development; policy differences on education. 'The new policies are more

likely to result in a greater concentration of scientific effort on problems of direct relevance to China's development needs.'

176 OLDHAM, C.H. Geoffrey
Science and education in China. *Contemporary China*, Ruth Adams (ed.). New York, Vintage Books, p. 281-317, 1966. Also *Bulletin of the Atomic Scientists*, 22(6), 41-50, 1966.
'This article is impressionistic and interpretive; it is not a survey of the field.' Education in China: Objectives; the education system in outline; higher education--academic work, politics, teaching staffs, material conditions; success in meeting objectives. Science in China: organization and development; observations on scientific research in certain institutes of the Academia Sinica; China and international scientific relations; popularization of science and technology; agricultural sciences in the communes; worker innovation in industry; science and society--present and future; Chinese development and world peace. Appendix: observations on certain institutes of the Academia Sinica.

177 OLDHAM, C.H. Geoffrey
Science and technology in China's future. *Contemporary China*, Toronto, The Canadian Institute of International Affairs, p. 113-29, 1968.
History of Chinese science; science in China during the Cultural Revolution. Science in China's future; 'the way in which science and technology contribute to--and indeed modify national goals, depends upon the socio-economic environment. What this environment will be after the death of Mao Tse-tung is impossible to foretell.' List of assumed goals and ways in which science and technology may be used to achieve them. 'The most intriguing unknown about science in China's future is the extent to which science and technology will modify China's national goals.'

178 OLDHAM, C.H. Geoffrey
Science for the masses? *Far Eastern Economic Review*, 60(20), 353-5, 1968.
One result of the Cultural Revolution has been an apparent change in priorities with more of China's scientists switched to working on projects of immediate relevance to China's development needs. This will be a loss to world science but may be a gain for China's development.

179 OLDHAM, C.H. Geoffrey
Science in China. *Aikya*, Student Christian Movement of India, P. 12-16, Mar 1970.
'At the start of the Cultural Revolution, China appeared to be making steady, but not spectacular progress toward catching up with the advanced countries as a major contributor to scientific knowledge.' Science since the Cultural Revolution: 'an objective appraisal of the overall status of science in China today is impossible.' 'So much depends on goals and objectives. If economic development is the sole objective the policies ascribed to Liu are probably correct. If total development is required, with perhaps, greater importance given to social and political

factors, then Mao's policies make good sense.' Mao Tse-tung favours 'self-reliance' for scientific and technological development.

180 OLDHAM, C.H. Geoffrey
Science in China's development. *Advancement of science*, 24(122), 481-7, 1968. Also *Far East trade and development*, 23(3), 223-7, 1968.
Text of a paper presented at the Leeds meeting of the British Association, 1 Sep 1967. Science in traditional China was of a pre-scientific revolution variety, where theory was not tested by experiment. History of the development of modern science in China. Science and national goals in the PRC. Manpower; organization of science; financial support for research. Appraisal of the science system: military research; science, technology and economic growth; science and social change. Impact of the Cultural Revolution.

181 OLDHAM, C.H. Geoffrey
Science in Mainland China: a tourist's impressions. *Science*, 147(3659), 706-14, 1965.
Report of a visit to China by British geophysicist in 1964. 'From the point of view of my scientific specialty of geophysics the trip was disappointing . . . . But from the point of view of the study of scientific development in Asia the visit was quite rewarding.'

181a *SCIENCE NEWS LETTER*
Scientific revolution brews in Red China. 87(9), 133, 1965.
Brief note on science in China based on article by C.H.G. Oldham in *Science*, 12 Feb 1965.

182 OLDHAM, C.H. Geoffrey
Science travels the Mao road. *Listener*, 22 Aug 1968. Also *Bulletin of the atomic scientists*, 25(2), 80-3, 1969.
Political disputes within the scientific establishment over research policy, application of science to agricultural and military goals; education.

183 OLDHAM, C.H. Geoffrey
*The scientific revolution and China*. Institute of Current World Affairs, unpublished newsletters CHGO-37, 2 Dec 1964.
Caryl Haskin's *The scientific revolution and world politics:* the social and political aspects of the scientific revolution. 'China is closer to achieving its scientific revolution than any of the other Asian countries I have visited. Worker--and peasant--inventors and 'scientists'. Spare-time schools. Popularization of science.

184 ORLEANS, Leo A.
China's science and technology: continuity and innovation. *People's Republic of China: an economic assessment*, U.S. Congress: Joint Economic Committee. Washington, U.S. Govt. Printing Office, p. 185-219, 1972. Adapted and abridged as How the Chinese scientist survives. *Science*, 177(4052), 864-6, 1972.
'The policies that guided scientific and technological development in China during the 15 years prior to the Cultural Revolution are

still in effect' . . . . Important changes brought by the Cultural Revolution were 'more a reflection of innovations in the economy and education than in science and technology per se.' 'Basically, then, China continues to 'walk on two legs' in science and technology.' Historical survey; continuity of ideology and policies; how the Chinese scientist survives; education and manpower for science and technology; science and technology in practice (R & D and innovation); publication and cross-pollination in science and technology (journals and books; exchange of information; international contacts).

185 ORLEANS, Leo A.
Research and development in Communist China: mood, management and measurement. *An economic profile of Mainland China*, U.S. Congress: Joint Economic Committee. New York, Praeger, p. 549-78, 1968. Also Research and development in Communist China. *Science*, Vol. 157, p. 392-400, 28 Jul 1967.
'I have tried to draw together some relevant information that will provide a panoramic view of the R & D picture, identify major information gaps, and provide a possible methodological approach for measuring Communist China's R & D effort.' Political and social setting; Soviet vs. Western influences; nature of research and development; structure, performers, content; measurement of R & D.

186 RESEARCH POLICY LIBRARY
Land and Research Survey and Planning Organization, New Delhi. *Research potential and science policy of the People's Republic of China; a bibliography*. New Delhi, Council of Scientific and Industrial Research, 1966.
'This bibliography aims to present the existing materials necessary for the study of China's research potential and research policy. The classification used is still at an early stage and only a temporary one.' Includes references to Soviet publications on Chinese science and science policy.

187 *SCIENTIFIC AMERICAN*
Science in Communist China. 204(2), 66-70, 1961.
Report of the AAAS symposium on Chinese science.

188 SHIH, Joseph Anderson
Science and technology in China. *Asian survey*, 12(8), 662-75, 1972.
'This article attempts to summarize the general condition of science and technology in the People's Republic of China during the past twenty years. From this, we may be able to forecast the future trend. And . . . we may be able to gain some understanding of the interplay of technology and ideology.' Based on refugee interviews in Hong Kong, travellers' reports and mainland newspapers and periodicals. General review of the Five-Year Plans; major achievements of Chinese science and technology since 1960. The reasons for the Chinese success: educational reform; utilization of foreign research; preferential treatment of scientists and technologists; competition and emulation campaigns; technological renovation and revolution. The future development of China's

science and technology: '. . . science and technology have reached a stage in China's closed society where they are recognized as having an important impact on social change. In these circumstances, technology threatens the dominance of ideology. And the institutions that control science and technology cannot avoid becoming targets of the power struggle of contending political factions.'

189 SIGNER, Ethan
New directions in Chinese science. *Science for the people*, 3(4), 1971. Adapted for *Technology Review*, 74(2), 8-9, 1971. Also *China now*, No. 20, p. 5-7, Mar 1972.
Report by the author, an American biologist, of his observations at various research institutes and universities on a visit to China in May 1971. As a consequence of the Cultural Revolution, the scientific establishment is being reorganized to integrate research more closely with the immediate requirements of industry and agriculture and to change social customs perpetuating a scientific and professional elite, while making scientists useful members of society. Reorganization of decision-making and administrative procedures in scientific institutions. Educational reforms. Quality of scientific research is modest, with significant exceptions; emphasis has shifted from basic to applied research. 'Walking on two legs': large scale science in research institutes, and research into applied problems on a small scale in factories and communes. '. . . the scientists we met . . . seemed remarkably and genuinely sympathetic to the egalitarianism and practicality that characterize the new science policy.'

190 SIGNER, Ethan and GALSTON, Arthur W.
Education and science in China. *Science*, 175(4017), 15-23, 1972.
Description by the authors of their visit to China in April and May 1971. 'We found that, under the impetus of the Cultural Revolution, the Chinese are experimenting with new ways of organizing science and medicine. They are trying to integrate scientific research more closely with the immediate needs of industry and agriculture, to broaden the scope of medical care so that it reaches as much of the population as possible, and to do away with institutional and social customs that used to keep intellectuals and professionals as elite classes culturally distinct from ordinary people.' University organization since the Cultural Revolution. Research: direction and administration; access to scientific publications; lines of research at the Biochemical Institute, Plant Physiological Institute, Botanical Institute, Microbiological Institute, various research laboratories of Chungsan (sic) University. '. . . the decision to concentrate on applied rather than basic research . . . often involves scientists working closely with a specific factory or agricultural commune.' 'Conversely, factory workers and peasants have begun spending a few weeks or months in the appropriate research laboratory to learn techniques.' 'Self'reliance' in local and national research. Attempts to alter the social framework of research: rotation of faculty members through teaching, research, and manual labor; more egalitarian structure of research groups. Medical care. General comments: 'The quality of most of the scientific research we saw was modest.' 'China has

obviously suffered a short-term loss in scientific productivity, but the Chinese feel that this will be more than compensated for by the long-term benefits that are expected to result from the reforms of the Cultural Revolution.'

191 SIGURDSON, Jon
*Naturvetenskap och teknik i Kina* (Natural science and technology in China). Stockholm, Swedish Academy of Engineering Sciences, 1968.
Survey of the literature on Chinese scientific and technical development in English, French and German; the author's experience as cultural attache in Peking. Research and industry organization: organization of scientific research; development and planning of research since 1949; basic research; education; research in institutions and ministries; organization of industry; cooperation between research and industry. Scientific documentation and information. Level and results of research. Bibliography. Appendix: list of Chinese newspapers and periodicals.

192 SIGURDSON, Jon
Vetenskapens roll i Kina (Chinese attitudes toward science). *Svensk Naturvetenskap*. Research Council for Natural Sciences, 1970.
Recognition of the importance of science and technology for modernization; scientific cooperation with Soviet Union in early 1950s. Foundations laid in Great Leap Forward for training of China's own scientific manpower and for R & D projects.

193 SLAVEK, P.
Utbildning och forskning i Kina. *Ekonomisk Revy*, Vol. 20, P. 60-2, 1962.
Brief article on education and research in China.

194 SNOW, Edgar
*The other side of the river*. New York, Random House, 1961. Also abridged as *Red China today*. Harmondsworth, Penguin Books, 1970.
Chap. 30--Science and education. Organization of science: the Chinese Academy of Sciences; China Federation of Scientific Societies; China Association for the Dissemination of Scientific and Technical Knowledge. Interview with Vice-Minister of Education, Tsui Chung-yuan: three stages in educational development in China since 1949.

195 SOREL, J.J.
La Chine nouvelle puissance scientifique. *Atomes*, No. 233, P. 341-5, Jun 1966. Also *Problemes economiques*, No. 983, P. 21-5, 3 Nov 1966.
Brief commentary on study of Chinese science policy, history of Chinese science. Post-1949 education of expert manpower. Female manpower. Planning and organization of science. Social sciences. 'A balance sheet.'

196 SURVEYS AND RESEARCH CORPORATION
*Directory of selected scientific institutions in Mainland China*. Stanford, Hoover Institution Press, for the National Science

Foundation, 1970.
'This directory represents a first major attempt to compile and publish information concerning research and development institutions in Mainland China. The fields of science covered are the physical, biological, medical and agricultural sciences, and engineering. Information is provided on 490 selected research and development institutions, including whenever possible the institution's location, year of establishment, facilities, nature of work, major activities, publications, and biographical sketches of key scientific or administrative personnel.' 'The classifications employed follow the structure of China's research and development industry . . .' 'I--Academies: Chinese Academy of Sciences; Chinese Academy of Medical Sciences; Chinese Academy of Agricultural Sciences. II--Other governmental agencies. III--Industrial enterprises. IV--Professional societies. V--Universities and specialized colleges: medical colleges; engineering and technical colleges; agricultural colleges; normal or teachers colleges; general universities.' Introduction includes Organization and management of science and technology.

197 SUTTMEIER, Richard P.
Scientific societies. Chapter in *Chinese scientific development*, East-West Technology and Development Institute, Working Paper No. 30. Honolulu, The East-West Center, Sep 1972.
'The argument running through this paper is that Chinese scientific societies formed an invaluable administrative resource for overcoming organizational deficiencies in a relatively differentiated and variegated system of science-related institutions.' Appendix: Professional societies in the natural sciences and engineering fields as of 1966.

198 THOMPSON, H.W.
Science in China. *International science and technology*, No. 18, P. 86-95, Jun 1963.
Report of a visit to China (1963?) by the author under Royal Society auspices. 'Broadly speaking, we found a scientific community which seems still in the process of building a good foundation. The Chinese laboratories we saw are learning the use of modern methods by repeating known observations, rather than aspiring to much that is really new.' Visits to Peking University, Tsing Hua Technological University, the Agricultural University of Peking. Description of research institutes of Academia Sinica and facilities.

199 TSIEN, Tche-hao
L'Enseignement superieur et la recherche scientifique en Republique populaire de Chine. (Higher Education and Scientific Research in the P.R.C.). Paris, Libraire Generale de Droit et de Jurisprudence, 1971. Excerpts as Une Conception Revolutionnaire de la Recherche. *La Nouvelle Chine*, No. 4, P. 27-30, Sep 1971.
I--General principles common to higher education and scientific research. II--Higher education: historical evolution; organization and functioning; The government organ: the Ministry of Higher Education, different types of higher education establishments, structure and organization of establishments, the system of

part-study/part-work, the socialist universities, parallel education; finance; personnel; studies; life of students. III--Scientific research: historical evolution; general organization of research; centers of research; research workers; some methodological and organizational principles. IV--Relations and conflicts between higher education and scientific research: general principles; relations between establishments; mobility of research workers and students.

200 DIE UMSCHAU IN WISSENSCHAFT UND TECHNIK
*Das Wissenschaftliche Leben in der Volksrepublik China*. 64(24), 741-4, 763-4, 1964.
Article based on Hans-Jürgen Eitner, 'Erziehung und Wissenschaft in der VR China'. Academic life in the PRC: historical development; all-round development to overcome the separation between mental and manual labor; organization of schools and universities and support of students; scientific academies; developments in the various fields of knowledge; mathematics, solid state physics, electrical engineering and electronics, engineering, chemistry and chemical engineering, geology, health and medicine.

201 UNION RESEARCH INSTITUTE
*Communist China problem research series*. Hong Kong, EC32--p. 199-224 Pai Chen, Scientific work, 1961. EC34--Vol. 2, P. 55-83, 1962, Communist China's natural science work in 1962. EC36--Vol. 1, P. 127-50, 1963, Wu Leng, Scientific development. EC37--Vol. 1, P. 172-210, 1964, Ch'iu Shih-chih, Scientific and technological development. EC39--Vol. 2, P. 208-29, 1965, Ch'iu Shih-chih, Communist China's work of scientific research in 1965. EC40--Vol. 2, P. 151-70, 1966, Ch'iu Shih-chih, Communist China's science work in 1966.

202 U.S. CONGRESS
*Senate Committee on Government Operations; Subcommitee on National Policy Machinery. National Policy Machinery in Communist China.* Washington, U.S. Government Printing Office, 1960.
Study of China's machinery for formulating and implementing national policies. Roles of the Party and the government; policy-making and execution in economic and scientific affairs. Section VI--Scientific Affairs, P. 26-8: organization of science policy institutions: The State Council; The Scientific and Technological Commission; The Academy of Sciences; The Academy of Agricultural Sciences; The Academy of Medical Sciences; ministerial research organs; institutions of higher learning. Party leadership in science policy and in individual research institutions. History of science policy; the 12-Year Plan.

203 *US NEWS AND WORLD REPORT*
'A threat to the United States in another 5 or 6 years', P. 52, 57. 14 Nov 1966.
Interview with C.Y. Cheng. China's missile development program. Other recent scientific achievements: synthesis of insulin. Speculation about effects of the Cultural Revolution on science.

204 WALSH, John
Manpower: Senate study describes how scientists fit into scheme of things in Red China, Soviet Union. *Science*, 141(3577), 253-5, 1963.
Based on U.S. Congress, Senate Committee of Government Operation, Subcommittee on National Security Staffing and Operations: *Staffing procedures and problems in Communist China* (entry no. 451) and Leo A. Orleans, *Professional manpower and education in Communist China* (entry no. 441); and on *Staffing procedures and problems in the Soviet Union* and Nicholas DeWitt, *Education and professional work* (sic) *in the U.S.S.R.*

205 WEDGWOOD BENN, Anthony
China--land of struggle, criticism and transformation. *New Scientist*, 53(777), 10-12, 1972.
The author records observations made on his trip to China in 1971. Effect of the Cultural Revolution on the organization of scientific institutions; elimination of duplication of effort by merging CAS institutes with industrial research institutes, or by decentralizing administration to provincial and local government level. Academy retains overall responsibility for research. Allocation of scientific resources to space, computer programs. Research workers sent to factories and countryside, then return to their institutes to work on practical problems; veteran workers lecture, bring problems to the institutes. Links between schools and universities and production. Reorganization of Chinese medicine follows the same patterns.

206 WILGRESS, D.
China's forward leap in science. *Discovery*, 21(11), 464-73, 1960.
The Twelve-Year Plan for science and technology; The Chinese Academy of Sciences stresses fundamental research; signs of increasing Chinese independence from Soviet scientific and technical support; education. China's advance in science will make it a great economic power; the West cannot afford to remain in ignorance of this development.

207 WILSON, Dick
*A quarter of mankind*. London, Weidenfeld, 1969. Also published as *Anatomy of China*. New York, Weybright and Talley, Inc. and Mentor Books, 1969.
Pt. 2: Economy. Chap. 10--Agriculture. 11--Population, birth control. 12--Industry. 'A start has been made, and there are industries in which China offers an example to other developing countries. The nuclear explosions are also a product of a well-prepared industrial effort. A question-mark, however, hangs over China's technology . . . can the Chinese Communist leaders be sure of safely 'going it alone' in this constantly changing world of industrial technology?.' 13--Science and technology. 'China has good scientists, and is giving them a reasonable budget, and yet doubts about the Communists' science programme are raised by the predominance of political considerations in its implementation.'

208 WILSON, J. Tuzo
*One Chinese Moon*. London, Michael Joseph Ltd., 1959.
The author, one of six Canadian delegates to the final meeting of the International Geophysical Year in Moscow, July-August 1959, describes his visit to China on the return journey. Visits to the Academy of Sciences in Nanking and Peking; to the Institute of Geology of the Academia Sinica; to the Central Geophysical Observatory of China; to Lanchow Geophysical Observatory; to the Institute of Geophysics and Meteorology of the Academia Sinica; to department of geology and geography, University of Peking; to meteorological and seismological observatory at Sian Northwest University; physics department of Peking University. Observations on libraries, instruments. Chinese participation in and withdrawal from International Geophysical Year.

209 WU, Yuan-li
Expansion of the Chinese research and development industry. *The China Mainland Review*, 1(2), 1-9, 1965.
The effective constraint to China's economic growth is the ability to convert resources freed from production of consumer goods to that of investment goods; this can be achieved through imports and through improvements in technology which alter the production function or the interrelationships between factors of production. Division by sectors and fields: growth of R & D in government, educational institutions, industrial enterprises and academies; co-ordination of R & D with economic policy--emphasis on applied research and development; half-hearted interest in basic research.

210 WU, Yuan-li and SHEEKS, Robert
*The organization and support of scientific research and development in Mainland China*. New York, Praeger, 1970.
'This study focuses on the acquisition and use of knowledge in science and technology through research and development in Mainland China . . . . The study seeks to establish the available facts on organization and support of science, and makes observations on the methods by which the Chinese have acquired and used new knowledge in the scientific and technological fields.' The study covers the life sciences, the physical sciences and mathematics, and the engineering sciences. 'Scientific activity' includes basic and applied research and development, testing and standardization, training, scientific and technical information activities, collection of general-purpose data, and popularization of scientific and technical information. Chap. II--History of China's science and technology to 1949. III--Formulation of goals and policy. IV--Trends in the structure of R & D. V--Financing R & D activities. VI--Users of funds for science. VII--R & D institutions in operation. VIII--Economic growth and the R & D effort.

210a BAUM, Richard
Chinese science. *Science*, 172(3984), 669-70, 1971.
Review of Yuan-li Wu and Robert B. Sheeks, *The organization and support of scientific research and development in Mainland China*. 'China's scientific establishment--its historical growth,

organization, training programs, financing, staffing, priority structure, and control apparatus--is the subject of this important and timely study . . . . Their findings help to place China's developmental dichotomy (the existence of sophisticated, modern scientific research and development within an essentially premodern, peasant-based society) in its proper perspective.'

211 YAMADA, Keiji
*Question for the future: The Chinese experiment.* Tokyo, Chikuma Bookstore, 1968.
Section III--Science and technology; outline of technology; the development of science and technology and its logic; labor, technology and humanity.

212 YANG, Chen Ning
Education and scientific research in China. In science and medicine in the People's Republic of China, *Asia,* No. 26, P. 74-84, Summer 1972.
Visits to three research institutes of the Chinese Academy of Sciences in 1971: Institutes of Biochemistry, Physiology and Nuclear Physics. Re-evaluation of research programs since the Cultural Revolution. Basic research will not have priority in education programs. Scientific publication and communication.

213 YOUNG, G.B.W.
Some remarks on scientific achievement in Communist China. *RAND Memorandum RM-3077-PR.* Santa Monica, The RAND Corporation, Mar 1962.
Monograph reviewing China's science budget; the initiation of the 'technical revolution' in 1958, indicating persistence of low-level unsophisticated technology, although efforts are also made to acquire modern technology; the development of mechanics in China as an example of scientific development. By 1967, China will have a firm base for achievement of its goals in science.

See also the following entries

Technology Policy:
  Technology and Economic Growth. 7, 20, 24.
  Innovation and the Direction of Technological Change. 69, 97, 99, 108, 113.
Policy Toward Science. 218-20, 225, 227, 243, 250.
Research and Development:
  Military/Space Exploration and Travel/Nuclear Energy. 317.
Higher Education and Technical Training. 412, 414, 433, 435, 439, 441, 448.
Industrial Technology:
  Electronics Industry. 516.
Medicine and Public Health. 713.

## B Policy toward Science

214 BARNETT, A. Doak
Mao versus Modernization. *Technology Review*, 71(1), 31-3, 1968.
'The trauma that China is now undergoing is related to conflicts in values which arise out of the processes and consequences of modernization.' Red vs. expert. The Cultural Revolution represents a significant setback for the modernization process--and for the development of science and technology in China.

215 BAUM, Richard D.
'Red and Expert': the politico-ideological foundations of China's Great Leap Forward. *Asian Survey*, 4(9), 1048-57, 1964.
The red expert campaign initiated in 1957 constituted a response to the lack of enthusiasm for socialism demonstrated by the 'old' intellectuals, and the lack of material (i.e. economic and technological) resources necessary to sustain rapid economic development.

216 CHANG, Alfred Zee
Scientists in Communist China. *Science*, 119(3101), 785-9, 1954.
Survey of leading Chinese scientists: general characteristics and present occupations. Communist policy and practice toward scientists. 'It is apparent that the Communists have little appreciation of the talents and ability of leading scientists.'

217 CHANG, Parris H.
China's scientists in the Cultural Revolution. *Bulletin of the atomic scientists*, 25(5), 19-20, 40, 1969. Also China's eclipse of the moon. *Far Eastern Economic Review*, 63(3), 97-9, 1969.
Immunity of the scientific community from political attack during the Cultural Revolution withdrawn in the early months of 1967.

218 *CHEMICAL & ENGINEERING NEWS*
China: science on a swinging pendulum. P. 17, 20-1, 6 Mar 1972.
Politics permeates all aspects of life in China . . . including science; the Cultural Revolution closed universities and slowed down research, but the pendulum now appears to be moving toward more latitude for scientists. The central role of science and engineering in China's development plans. Little basic research; science is seen as a collective group activity, uniting theory and practice to serve production. Stress on self-reliance, 'walking on two legs'. 'It appears that agriculture and military research and development are the top two priorities, with medicine not far behind them.' Chinese efforts in the environmental field. The problem of training a scientific elite capable of independently conducting and supervising research; the cost to scientific capabilities in neglecting basic research.

219 CHEN, Theodore H.E.
Science, scientists and politics. Sciences in Communist China, Gould. *AAAS* No. 68, P. 59-102, 1961. Excerpts in *Foreign Policy Bulletin*, Vol. 40, P. 81-2, 86, 15 Feb 1961.

Thought reform of Chinese scientists. Scientists in the Hundred Flowers Movement and subsequent denunciation of rightists. New points of emphasis: party leadership; science must take the 'socialist road'; integration of science with production; centralized planning of science and collective work; scientists must be 'red-experts'.

220 CHEN, Theodore H.E.
*Thought reform of the Chinese intellectuals.* Hong Kong, Hong Kong University Press, 1960.
Twelve-Year Plan for the development of science and technology, p. 115, 116. Greater material incentives offered scientists than other intellectuals, p. 190. Thought reform of scientists, p. 36. Party interference in scientific research, p. 96, 97. 'Bourgeois ideology'--'Technical viewpoint'; preoccupation with technique, p. 55.

221 *CHINA NEWS ANALYSIS*
Ch'en Yi on 'Red and Expert'. No. 393, 20 Oct 1961.

222 *CHINA NEWS ANALYSIS*
Chinese experts. No. 458, Mar 1963.
Brains needed: *People's Daily* editorial on 'combination of three', 11 Feb 1963. Scientists in 1962. Note: list of 39 reviews of natural science.

223 *CHINA NEWS ANALYSIS*
Life in the Academy of Science. No. 843, 4 Jun 1971.
Review of research and political consequences for the CAS during the Cultural Revolution: review of press items 1968-1971, annually. 'This is all that we have been able to collect from three years' reports about the Academy of science and scientific research. What is said about scientific institutes and scientists, what they are doing and how they are being treated, sounds so odd as to be almost unbelievable. The Academy of Science is under the dictation of ignorant soldiers. Specialization and theoretical studies are being discouraged. Some of the reports quoted express what the scientists themselves think about this, how discouraged they are, and how difficult they find it to adjust themselves to this new world.' (See also Deborah Shapley in *Science*, 13 Aug 1971, entry no. 251).

224 *CHINA NEWS ANALYSIS*
Revival of science 1961. No. 385, 18 Aug 1961.
With the collapse of food production at the end of 1960, the Party made concessions in science and retreated from extreme anti-science position of the Great Leap. Debate on 'specialized science' and 'specialized trade science' proposed by Mao I-sheng.

225 *CHINA NEWS ANALYSIS*
Scientists, 1957-1967. No. 696, 16 Feb 1968. Trans. as La Recherche Scientifique (1957-1967). Problemes Chinois No. 5, *Notes et Etudes Documentaires*, Nos. 3498-3499, P. 59-65, 15 Jun 1968.
Denigration of science during Great Leap; concessions made to scientists after 1960. Policies for promoting pure science, basic

research, 1964; growing insistence on applied research for industry, 1965 but a compromise on science policy prevailed. 1966: the Cultural Revolution turns against science. 1967: personnel changes in the science establishment.

226 *CHINA REPORT*
The father of Chinese rocketry, 3(1), 38-40, 1966-7.
Biographical note on Ch'ien Hsueh-shen.

227 *CHINA REPORT*
Two demands on science, 1(5), 3-8, 1965.
Organization of science. Scientists in the Hundred Flowers period; thought reform of scientists; redness and expertness; integration of theory and practice; 'triple combination'.

228 CHOUARD, Pierre
La recherche scientifique ou lendemain de la Revolution Culturelle (Scientific research on the morrow of the Cultural Revolution). *La Recherche*, 3(23), 411-16, 1972.
The author's observations of the effect of the Cultural Revolution on Chinese science. Two practical effects: scientists are 're-educated' by spending two years living in peasant villages or with workers; and the experience of workers and peasants is a source of scientific inspiration. The author's visit to the Nanking Observatory. The 're-education' of a geneticist, who dropped his research on genetic theory in favor of applied research on hybrid rice strains. The Cultural Revolution emphasizes applied research: for example, research on potatoes at the Institute of Genetics. At the Photosynthesis Laboratory of the Institute of Plant Physiology of the CAS: 'pure science', rigorously conducted according to the concepts of science, and 'fundamental science' conducted with the same experimental rigor, but for the purpose of innovation. Botany.

229 CHRISTIANSEN, Wilbur N.
A foreign scientist in the Chinese Cultural Revolution. *Wetenschapen en Samenleving*, 24(6), 169-75, 1970.
'I shall try to stress the difference between the attitudes of Chinese and Western scientists to their work and the difference in the way in which they tackle scientific tasks.' General background to the Cultural Revolution. The author's experiences working as a radio-astronomer in China, 1966-67. Background to China's resignation from the International Astronomical Union in 1961.

230 CHRISTIANSEN, Wilbur N.
Science and scientists in China today. *Scientific research*, P. 64-72, Oct 1967.
Observations by Australian electrical engineer in China, 1966-67, as adviser on construction of a radiotelescope. Scientific work in China will not develop along the same lines as in the West; science in China will be considered in relation to the progress and future life of the people of the world--'politics comes first'; pre-eminence of man over machine, reform of scientists. Educational reforms. Influence of Soviet education and research, followed by Chinese attainment of self-sufficiency.

231 CLAUSER, H.R.
China's research becoming visible again. *Research Management*, 14(5), 4-5, 1971.
Brief report based on M. Macioti (*Science Policy News*, entry no. 158, May 1971): '(China's recent Cultural Revolution) has sought to purge Chinese science and technology of elitism and to reshape it towards an egalitarian society, in which the research endeavor is shared by all.'

232 DOOLIN, Dennis J.
'Both Red and Expert': The dilemma of the Chinese intellectual. *Current Scene*, 2(19), 1963.

233 *THE ECONOMIST*
Mao calls the boffins home, Vol. 179, P. 1175-6, 23 Jun 1956.
One reason for the Twelve-Year Plan for the development of science and technology is 'the failure of the Communists, despite strenuous efforts, to attract home' ex-patriate Chinese scientists. Relaxation of controls on intellectuals in order to induce return of scientists.

234 ESPOSITO, Bruce J.
*The Cultural Revolution and science policy and development in Mainland China*. Canberra, paper presented at the 28th International Congress of Orientalists, Jan 1971.
. . . 'The Cultural Revolution has intensified the emphasis placed on production-related research . . . . Scientists themselves have been the targets of purges and re-education drives aimed at increasing their political awareness . . . the advancement of scientific knowledge in China has been significantly retarded . . . . The short-term effect of the educational 'reform' would appear to be a shortage of competently trained research aides and researchers. The long-term effects . . . could severely limit the number of competently trained middle and top echelon scientists . . . if the Chinese xenophobic attitude finds fertile ground in the field of science, then Chinese science will certainly be retarded in development. 'In sum the Cultural Revolution has had a detrimental effect on Chinese scientific development. Some of this effect could be remedied in time by changes in policy, something the Mao-Lin leadership appears to be incapable of making.'

235 ESPOSITO, Bruce J.
Science in Mainland China. *Bulletin of the atomic scientist*, 28(1), 36-40, 1972.
The Cultural Revolution disrupted promising scientific developments underway in China in 1966. Résumé of events in scientific institutions from 1966, based on 'big character posters', *Red Flag*, *People's Daily*, monitored radio broadcasts, and U.S. government translation services. 'It would appear that the Cultural Revolution has exacerbated the hostility between the scientist and the Communist Party cadre.' 'The political leadership appears to have been relatively successful in insuring that research projects are directly related to either current or

future production . . . this policy rigidly applied can further widen the fissure between the natural instincts of the scientists and the Party.' Science may be retarded by Chinese xenophobia. The effect of the Cultural Revolution on military research.

236 *KONTAKT*
De Kinesiske Forskere Skal Tjene Kina--Ikke Sig Selv. 21(7), 43-4, 1968.
The Chinese researcher must serve China--not himself. Conversation with Stefan Dedijer: effect of the Cultural Revolution on science in China.

237 LEE, Renssalaer W. III
The 'Hsia Fang' system: Marxism and modernization. *The China Quarterly*, No. 28, P. 40-62, 1966.
History of the Chinese Communist Party's policy of manual labor for intellectuals.

238 MACDOUGALL, Colina
The Reds and the experts. *Far Eastern Economic Review*, 43(6), 310-12, 1964.
Loss of Soviet technical aid made China more than ever dependent on scientists trained before 1949 and has resulted in an irreversible trend toward liberalized treatment of experts.

239 MACFARQUHAR, Roderick
*The Hundred Flowers Campaign and the Chinese intellectuals.* London, Stevens and Sons Ltd., 1960.
Translations of criticisms made by the Chinese intelligentsia in 1957. Chap. 6--Scholars, university teachers. P. 82ff, The Academy of Sciences; universities; Ministry of Geology; Central China Engineering Institute. P. 112, The Democratic League's academic programme. Chap. 7--Doctors.

240 NEEDHAM, Joseph and YAMADA, Keiji
Science in China. *Mainichi Daily News*, Tokyo, 17, 18, 19 Jul 1971.
The 'gist of a dialogue' held in Kyoto. The meaning of the Cultural Revolution for Chinese science. Specialized training is necessary for the development of science and technology, but a strong ethical motivation may keep the ties between scientists and the masses. Traditional and modern science and technology cannot be mixed in all fields, although steel refining, medicine are examples of successful mixing. Needham's observations of research on the neuro-physiological aspects of acupuncture. Chinese science and technology will not be different from world science, although there are some 'Chinese' characteristics: yin-yang and the 'organic' quality of Chinese scientific thinking. Chinese realize that science is universal and international and do not want to be isolated.

241 *NEW SCIENTIST*
Turning the heat on Chinese scientists. 'Notes on the news', 39(607), 169, 1968.
Brief review of report in *Scientific Research* of political attacks on Chinese scientists, 24 Jun 1968.

242 OLDHAM, C.H. Geoffrey
*China in retrospect.* Institute of Current World Affairs, unpublished newsletter CHGO-47, 12 Mar 1965.
Scientist: politician relations have followed a cyclic path of more or less government political interference in science affairs, ever since the communist victory in 1949.

243 *PROGRES SCIENTIFIQUE*
Carriere et mobilite des chercheurs en Chine. (Career and mobility of researchers in China). P. 38-40, Nov 1969.
The notion of a 'scientific career', after the Soviet system, changed to 'service of the people' in the Cultural Revolution. Mobility of researchers prior to the Cultural Revolution within the scientific establishment: organization of research into the CAS, higher education sector, technical ministries, military research.

244 PRYBYLA, Jan S.
Communist China's strategy of economic development 1961-1966. *Asian Survey,* 6(11), 589-603, 1966.
Movement to send educated youths to the countryside to prevent China's age-old problem of social alienation of the intellectuals from the laboring masses; 'combination of the three' to substitute for foreign scientific and technical assistance; scientific and technical education; self-reliance.

245 RAY, Denis
'Red and Expert' and China's Cultural Revolution. *Pacific Affairs,* 43(1), 22-33, 1970.
In the Cultural Revolution, the reds have been purged while the experts, on the whole, have thrived.

246 RYAN, William L. and SUMMERLIN, Sam
*The China Cloud.* London, Hutchinson & Co., Ltd., 1969.
Popularized reconstruction of the career of Tsien Hsue-shen and other Chinese scientists trained in Europe and the United States. Chinese nuclear and rocketry programmes in the context of national and international politics. Appendix: List of Chinese scientists with European and American training working on bomb and rocket research; eighty American-trained Chinese scientists, listed by American institution, attended.

247 SAITO, Akio
The great Cultural Revolution and the movement to study philosophy. *Monthly Bulletin of China Research.* Tokyo, China Research Institute, No. 280, Jun 1971.
Chap. I--Science, technology and the mass movement, p. 1-6.

248 SALAFF, Stephen
A biography of Hua Lo-keng. *Isis,* 63(217), 143-83, 1972.
Biography of the political and professional career of Chinese mathematician Hua Lo-keng (b. 1910): 'an indispensable resource person for those concerned with China's higher educational development . . . to gain insight into Chinese contributions to contemporary mathematics, and to learn how modern science has fared in China.

one must become acquainted with the life of Hua Lo-keng.' 'Hua's scientific and political roles cannot be uncoupled . . . .' 'This biography represents, in the first place, an appreciation of his scholarly contributions and of his literate, versatile mathematical style.' Hua's political reputation '(though not his mathematical research career)' was enhanced after his self-confession during the Cultural Revolution. Hua Lo-keng espoused changes in Chinese educational policy developed since 1969, which do 'not fully accommodate the mathematical creativity of his colleagues and students. The consequences within China are yet to be determined, but there has been a loss to world science and culture as a whole.'

249 *SCIENCE NEWS LETTER*
Red China's Michurinism. Vol. 66, P. 22, 10 Jul 1954.
Brief article based on report by Dr. Alfred Zee Chang (entry No. 216). 'The Communists worship the Russian horticulturalist (Michurin) in a way the Chinese used to do honor to Confucius.' 'It is understandable in the Western world that when Michurinists are dominant, the usefulness of modern scientists is doomed.' 'Michurin is . . . the symbol of the value placed on practical applications of plant breeding to benefit the state as opposed to "pure science" research.' Political attitudes toward scientists.

250 *SCIENTIFIC RESEARCH*
Two kinds of scientists in China: Maoists do well. No. 2, P. 71-2, Oct 1967.
Article based on interviews with Japanese scientists returned from China. Young, politically-motivated Chinese scientists are replacing senior research scientists in supervisory positions. The institutional infrastructure; concentration of effort on physical and chemical sciences, relative neglect of research in life sciences.

251 SHAPLEY, Deborah
Chinese science: what the China watchers watch. *Science*, 173(3997), 615-17, 1971.
'. . . there are many different perspectives on what is happening in Chinese science and to Chinese scientists.' Criticism of 'Life in the Academy of Science', *CNA* no. 843, 4 Jun 1971, (entry no. 223) based on an interview with Leo Orleans.

252 SHEEKS, Robert B.
*Science, technology, and the Cultural Revolution in China*. China, Columbia University Seminar on Modern East Asia, 11 Oct 1967.
In the long run the predominant characteristics of the Maoist Cultural Revolution cannot help but have a stultifying influence on China's drive for scientific modernization. Outline of suggested questions for further study of Chinese science and technology.

253 SHIH, Vincent Y.C.
The state of the intellectuals. *Communist China, 1949-1969: A Twenty-Year Appraisal*, Frank N. Trager and William Henderson (eds.). New York, New York University Press, Chap. 10, 1970.

P. 232-237: The fate of scientists and writers. Review of campaigns to indoctrinate scientists in principles of the Communist Party. Details of individual scientists' experiences during the Cultural Revolution. 'Although the Communist Communists, before the Hundred Flowers period, reluctantly acknowledged the classless nature of science so as to induce scientists to join the movement . . . science . . . is a handmaid of politics, and . . . technical expertise should be subordinate to 'redness'. Theory itself has no value unless it is employed in the service of production under a master plan worked out by the state.'

254 SUTTMEIER, R.P.
Party views of science: the record from the first decade. *The China Quarterly*, No. 44, P. 146-68, Oct/Dec 1970.
'This article attempts to elucidate the CCP's perceptions of attitudes towards science, which underlie the CCP's actions vis-a-vis the Chinese scientific community.' Two themes: 'First, the CCP's purposes for scientific activity are clearly utilitarian.' 'Second, the CCP's understanding of the past development of science and of how to foster continued scientific progress is radically sociological.' The impossibility of realizing all elements of the CCP views simultaneously and of simultaneously maximizing the utilitarian values attached to science has 'resulted in a fragmented science system of dissimilar institutions.' Implementation of CCP science doctrine: realization that a long-term process of resocializing senior scientists and creation of a new research tradition are necessary to 'make science serve production'; realization that social transformation would not result from organized 'institutional' research, thus the attempt to promote 'mass science'. Efforts to 'reconcile mass science activities with more conventional professionalized research . . . are a continuing challenge to China's science administrators.'

255 *US NEWS AND WORLD REPORT*
A look at science in Red China. Vol. 45, P. 107, 7 Nov 1958.
Interview conducted in Taipei with Dr. Wang Chi-hsiang, Chinese chemist trained in the U.S., who returned to China in 1957 but left after five months.

256 WILSON, Dick
Technology in China. *Far Eastern Economic Review*, 50(6), 289-91, 1965.
There are reservations about the quality of China's self reliant scientific technology: inhibition of the development of a sound modern science by the Communist Party's anti-intellectual and anti-foreign prejudices; prejudice against pure science and research; survival of traditional unscientific attitudes.

257 YAMADA, Keiji
Revolution culturelle et tradition chinoise. *Esprit*, P. 867-99, Dec 1971.
Translation of the author's paper presented at a conference on China in Tokyo, Oct 1969. Tradition and modernity are not mutually exclusive, but the active principle at the core of tradition should be retained to assure a fundamental continuity

in the development of the society and culture. In a given culture or society, it is the value system which decides the role and function of science and technology, and thus the 'active' or 'internal' principles of its tradition acts on the orientation of the modernizing society. Two propositions, illustrated by Chinese examples: a value only revives by changing itself; and, thought patterns have a strong tendency to become permanent. Together, the 'native' method and the triple combination constitute an implicit criticism of the epistemological method on which, it is presupposed, modern science is based. By affirming that human creativity derives from practice and that practice inspires man far more than theory, it lays the basis for a new technological and scientific system and attacks one of the main causes of our alienation.

See also the following entries:

Technology Policy:
  Technology and Economic Growth. 21.
  Innovation and the Direction of Technological Change. 72, 73, 99.
Policy for Science. 117, 119, 130, 132, 135, 138, 144-6, 150, 152, 155, 166, 167, 184, 185, 188, 190, 202, 203, 207.
Research and Development:
  Mathematics. 258.
  Bctany. 288
  Geophysics. 299.
  Military/Space Exploration and Travel/Nuclear Energy. 344, 354.

# SECTION IV

## *Scientific Activities*

There is a substantial body of literature reporting and commenting on research and other scientific activities in China; from this material, it is possible to infer Chinese science policy. The purpose in this section of the bibliography is to facilitate comparisons with the scientific activities of other countries. To this end, a modified version of the OECD scheme has been used for classifying research and development and activities related to or supporting R & D.

The Organization for Economic Co-operation and Development has attempted to standardize definition of scientific activities in its publication *The Measurement of Scientific and Technical Activities: Proposed Practice for Surveys of Research and Experimental Development* (the 'Frascati Manual', DAS/SPR/70.40., 1970), taking note of the earlier efforts by other organizations—specifically UNESCO and the Council for Mutual Economic Assistance—to standardize international practice. The Frascati definitions of scientific activities are used here as a basis for classifying items on scientific activities in China; this scheme of classification and the adaptations made for the purposes of this bibliography, are described below.

A *Research and Development*

The characteristic activity of modern science that is the most clearly identified with technological innovation is R & D—research and (experimental) development. The activities included in the OECD definition of R & D are

1 Basic research: 'original investigation undertaken in order to gain new scientific knowledge and understanding, not primarily directed towards any specific practical aim or application.' Basic research may be pure basic research, undertaken for its scientific interest, or oriented basic research, directed toward a field of 'present or potential scientific, economic, or social interest'.

2 Applied research: 'original investigation undertaken in order to gain new scientific or technical knowledge, directed primarily towards a specific practical aim or objective'.

3 Experimental development of 'new or substantially improved materials, devices, products, processes, systems, or services'.

'Special borderline cases' between experimental development and production and technical services, also included in the OECD definition of R & D, are

4 Design, construction and testing of prototypes.

5 Construction and operation of pilot plants.

Descriptions of these activities in China are all classified in the section on R & D. The OECD definition excluded 'trial production, trouble-shooting and engineering follow-through', except for 'feedback R & D'.

The Frascati Manual emphasizes the difficulty of distinguishing among these activities in practice, however, and proposes the classification of the full range of R & D activities according to the economic sector in which institutions perform or finance these activities. The Manual proposes four such sectors, namely the business enterprise sector; the government sector; the private non-profit sector; and the higher education sector. Given the organization of economic, political and other social institutions in China, this scheme is not appropriate to the classification of R & D in this bibliography. For example, research in the 'business enterprise', or industrial sector, and in the higher education sector in China, is subject to direct government policy. There is probably no equivalent in China to the 'private, non-profit' sector. It may be noted however, that in China research and experimental development may occur informally as non-specialized activities outside of the research institutes of the Chinese Academy of Sciences, or the research facilities in industrial enterprises, various government ministries, and in the universities. However, R & D of an unconventional nature—for example, agricultural 'research' carried out by peasants and industrial 'worker innovation'—has not received adequate attention in the secondary literature nor has its significance for technological advance and economic and social development been fully analysed.

As an alternative, the Frascati Manual recommends classifying R & D in the government sector according to the 'field of science'. The OECD scheme adopts the UNESCO definition of scientific fields: (A) Natural Sciences; (A1) Mathematics); (A2) Physics, Astronomy, Electronics, Mechanics; (A3) Chemistry and Physical Chemistry; (A4) Biology, Botany, Zoology, Bio-Chemistry, Bio-Physics; (A5) Geology and Earth Sciences, Meteorology,

Geophysics.[1] General descriptions of academic research have been classified under these headings in this bibliography.

In addition, the Frascati Manual suggests that R & D may be classified according to 'socio-economic objective'. Some of the literature on research and experimental development in China lends itself most readily to classification in these terms. Of the twelve such objectives proposed, (A) Military, (B) Space Exploration and Travel, and (C) Nuclear Energy have been combined here as section 5 in the R & D category. For (D) Agriculture, (H) Health, and (K) Industry, see sections 3, 4 and 2, respectively, under R & D, where cross references are supplied to appropriate entries in Section V on 'Technology in China', dealing *inter alia* with research and experimental development directed toward those purposes.[2]

The secondary literature which describes scientific activities in China focuses chiefly on research. These items range from evaluations of the quality and level of scientific research, to descriptions of research programmes, procedure, institutions and the organization of research. Scientific papers, the end product of research, are not cited in this bibliography.

Scientific research and experimental development requires the performance of certain supporting technical services.

B *General Purpose Data Collection*

This section involves non-research activities of museums, botanical and zoological gardens and nature reserves; technical survey work; and resource survey activity (social and economic data collection and analysis).

C *Scientific Communication*

This section includes the collecting, coding, recording, classifying, disseminating and translating of scientific and technical information concerning personnel, bibliographic services, patent services, official scientific and technical information services, and scientific conferences.

1 The continuation of the OECD/UNESCO classification by field of science is (B) Engineering; (C) Medical Science; (D) Agriculture; (E) Social Sciences; (F) Humanities and Fine Arts.

2 Other socio-economic objectives of R & D listed in the OECD scheme include (E) Construction; (F) Transportation; (G) Telecommunications; (I) Natural Environment; and (J) Under-developed Regions. The secondary literature does not discuss research directed to these ends, but see the relevant sections of 'Technology in China'. For (L) Academic Research, consult the section of R & D by field of science (section 1).

D *Higher Education and Technical Training*

This is one of the most important activities of the scientific community: essentially, the perpetuation of itself through creation of succeeding generations of skilled manpower. This education and training generally takes place in universities and polytechnical colleges with widely varying patterns of organization. In China, these institutions have been the subject of considerable organizational experimentation.

Scientific education does not begin at the tertiary level, of course; the teaching of science may begin in secondary and even in primary school. In addition, some students of science policy and development regard modern science as an agent of social and cultural, as well as economic, change in the developing countries. For this reason, the activities described in

E *Scientific and Technical Education at the Primary and Secondary Levels; Extension of Scientific and Technical Information; Popularization of Science*

are included in this bibliography as activities of the Chinese science systems.

'Research' is characterized by the discovery of new knowledge, and both the OECD and UNESCO classification distinguish R & D from other scientific activities by 'the presence . . . of an appreciable element of novelty'. New technologies are the product of research and experimental development, but the performance of R & D related, routine scientific and technological services is also important in the use of modern technologies, and indirectly in the promotion of technological advance.

F *Testing and Standardization*

This section includes feasibility studies for engineering projects; design and engineering services; specialized medical care; patent and license work; national patent office and government licensing activity; technical and scientific advisory and consultancy services.

Thus, the categories used to classify the literature in this section are

A. Research and Development (R & D)
   1. The Natural Sciences: Mathematics; Physics, Astronomy, Electronics; Chemistry; Biology, Botany, Zoology, Biochemistry; Geology and Earth Sciences, Meteorology, Geophysics
   2. Industrial R & D
   3. Agricultural R & D
   4. Medical R & D

5. Military, Space Exploration and Travel, and Nuclear Energy R & D

B. General Purpose Data Collection
C. Scientific Communication
D. Higher Education and Technical Training
E. Scientific and Technical Education at the Primary and Secondary Levels; Extension of Scientific and Technical Information; Popularization of Science
F. Testing and Standardization.

# A Research and Development (R & D)

## *1 The Natural Sciences*

### Mathematics

258 DAVIS, Chandler
A mathematical visit to China. *Canadian Mathematical Conference Bulletin,* P. 2-3, 5, 7-8, Fall 1971.
Based on the author's 1971 visit to the Mathematics Institute (CAS?), Peking University, and Futan University. Experiences of individual Chinese mathematicians during the Cultural Revolution. Research in applied mathematics predominates. University teaching of mathematics, and popularization of mathematical techniques, such as linear programming or critical-path method. There is nothing corresponding to graduate education in mathematics. Preparation for a 'resurgence of theory' by writing 'summaries' of mathematics. Publication. Western mathematical journals are received and used.

259 JAMES, I.M.
Visit to China. *The Royal Society,* London, Publication O/9 (67), 1967.
Observations from a 1966 visit to China. Contents include: The Academia Sinica and other bodies; mathematicians in Peking and outside Peking; notes on the Cultural Revolution; published mathematical research in China.

260 STONE, Marshall H.
Mathematics, 1949-1960. Sciences in Communist China, Gould. *AAAS* No. 68, P. 617-30, 1961.
'Although mathematics is more actively pursued than ever before in China, it is still largely derivative, is still rather concentrated on the more detailed aspects of classical problems, and is still almost entirely dependent for its scientific inspiration upon the leadership of a few gifted Chinese mathematicians and politically controlled contacts with the mathematicians of other Communist countries, chiefly the Soviet Union.

See also the following entries:

Policy for Science. 131, 149, 200.
Policy Toward Science. 248.
Scientific Communication. 378, 392, 393.
Higher Education and Technical Training. 449.

## Physics

261 BEYER, Robert T.
Solid state physics. Sciences in Communist China, Gould. *AAAS* No. 68, P. 645-58, 1961.
'In many respects, physics in China today is comparable to physics in the Soviet Union in the 1930s.' An attempt to survey all non-nuclear physics in China.

262 BURHOP, E.H.S.
Physics in Modern China. *Nature*, 178(4526), 184-5, 1956.
Observations based on a 'recent visit' to universities and scientific institutes in China. Ambitious expansion of physics training program does not seem to have lowered standard of courses. Curriculum of new five-year course, based on Soviet system, adapted to Chinese traditions. Some research in solid-state physics and crystallography in universities. Apparatus used in Nuclear Physics section of Physics Institute of Academia Sinica.

263 KURTI, N.
Notes on a visit to China. *The Royal Society*. London, Publication 0/5 (65), 1965.
Observations from a 1964 trip to China, including visits to Research Institute of Physics, Shanghai and Hangchow.

264 G.B.L.
C.N. Yang discusses physics in People's Republic of China. *Physics Today*, 24(11), 61-3, 1971.
Report of a visit to China by Chen Ning Yang in summer 1971. Institutions visited include Futan University; Institutes of Biochemistry and Physiology, Shanghai; Peking University; Tsinghua University; Institute of Nuclear Physics. Physics in China: development of a theory of 'strations' similar to quark theory; nuclear and solid state research; plans for high-energy and heavy-ion accelerators. Production of sophisticated industrial goods, e.g. 125.M W generators, hydraulic presses. Physicist Chang Wen-yu, deputy chairman of Revolutionary Committee of Institute for Nuclear Physics, directs scientific program. Physics at Peking University emphasizes semi-conductor work, cooperates with factories. American scientific publications received with time delay of 3 or 4 months. 'This attitude of making contributions to mankind is one reason for supporting such esoteric fields as high-energy physics. But another justification, Yang noted, is the technological fallout from scientific research' . . . e.g. synthesis of insulin led to capability to manufacture chemicals, especially enzymes.

265 LUBKIN, Gloria B.
Physics in China. *Physics Today*, 25(12), 23-8, 1972.
Summary of reports from seven U.S. physicists who 'recently' visited China. Organization, research programs, equipment, as observed at: Institute of Physics, Chinese Academy of Sciences; Peking University; Tsinghua University; Futan University; Shanghai Industrial Exhibit; Institute for Computer Research; Institute of

Semiconductors, CAS; Institute of Electronics, CAS; Miyun Observatory; Purple Mountain Observatory; University of Nanking; Mechanics Research Institute, CAS; Institute of Nuclear Physics, CAS. Research in astronomy and fluid mechanics; nuclear physics; high-energy physics. Scientists interviewed include Marvin Goldberger, C.K. Jen, Rudolph Hwa, Raphael Tsu, Chen Ning Yang, C.C. Lin, Chang-Yun Fan.

266 MIKHAILOV, I.G.
Ultrasonics in the Chinese People's Republic. *Akusticheskii Zhurnal*, 6(1), 139-41, 1960.
Research program in ultrasonics as described by author on basis of 3½ months at Nanking University, where he lectured on molecular acoustics and acted as consultant in setting up research program in the acoustics faculty.

267 OLIPHANT, Mark
Over pots of tea: excerpts from a diary of a visit to China. *Bulletin of the atomic scientists*, 22(5), 36-43, 1966.
Report of a visit to China in 1964 by Australian physicist.

268 WU, T.Y.
Nuclear physics. Sciences in Communist China, Gould. *AAAS* No. 68, P. 631-43, 1961.
Survey of research in nuclear physics: specific topics and results in experimental technology, nuclear structure, and elementary particles.

See also the following entries:

Technology Policy:
  Innovation and the Direction of Technological Change. 97.
Policy for Science. 131, 144, 149, 173, 200, 208, 212, 213.
Scientific Communication. 378, 393.
Higher Education. 453.
Industrial Technology:
  Electronics Industry. 516.

## Astronomy

269 CHRISTIANSEN, W.N.
A radio astronomer visits China. *Eastern Horizon*, 3(12), 33-5, 1964.
Observations from the author's visit to Peking Observatory and Nanking Purple Mountain Observatory. Education of astronomers.

270 WOOD, Frank Bradshaw
Astronomy. Sciences in Communist China, Gould. *AAAS* No. 68, P. 671-83, 1961.
Publications; observatories; departments of astronomy; collaboration with other nations; satellites.

See also the following entries:

Policy for Science. 149, 173, 208.
Policy Toward Science. 228, 229, 230.
Research and Development:
  Physics. 265.
Scientific Communication. 378.

## Electronics

271 *DATAMATION*
China has some computer building, P. 94-5, Jul 1972. Also *Soviet Cybernetics Review*, 2(5), 5, 1972.
Visit of Chinese computer specialists to Canada. Computers are designed and built in China; both French and English computers have been purchased. 1000 scientists, engineers, technicians, students, and workers are affiliated with the Institute of Computing Technology, which has 10 divisions and a factory.

272 LI, Yao-tzu and WOO, Way Dong
Progress in electronics 1949-1959. Sciences in Communist China, Gould. *AAAS* No. 68, P. 739-46, 1961.
Areas of electronics surveyed are radio communication, analog and digital computer technology, instrumentation and automation. Progress in these fields has been slow in China but remarkable progress made in the late 1950s.

See also the following entries:

Research and Development:
  Physics. 265.
Scientific Communication. 378.
Industrial Technology:
  Electronics Industry. 507, 512, 513, 516, 518.

## Chemistry

273 MARTIN, David
China today. *Chemistry in Britain*, 8(12), 533, 1972.
Brief note on the visit of the executive secretary of the Royal Society to the Academia Sinica: Institute of Chemistry, and to Peking National University, including the biochemical department, in May 1972. 'The visit to the Institute of Chemistry revealed the same pattern as elsewhere--a revolutionary committee in charge and a program devoted to applied work . . . close cooperation between the Institute and the factories with frequent interchange of personnel.'

274 WAY, E. Leong
Pharmacology. Sciences in Communist China, Gould. *AAAS* No. 68, P. 363-82, 1961.

The emphasis on pharmacology in Communist China is probably greater than in any other country. Pharmacology is used to exploit traditional Chinese medicine. Influence of pharmacologists trained in the West is greater than Soviet influence. Emphasis is on developmental rather than fundamental research; basic research has not been made academically significant contributions.

275 YU, Arthur
Chemistry. Sciences in Communist China, Gould. *AAAS* No. 68, P. 659-70, 1961.
Survey of research in chemistry based on various Chinese chemical publications available in the Western world. Organic chemistry; physical chemistry; inorganic and analytic chemistry; biochemistry; polymer chemistry.

See also the following entries:

Technology Policy:
  Innovation and the Direction of Technological Change. 97.
Policy for Science. 131, 149, 156, 200.
Scientific Communication. 370, 374, 378, 393.
Medicine and Public Health. 685, 696, 714.

## Biology

276 *CHEMISTRY*
Molecular biology in China. 41(9), 30, 1968.
Brief summary of review of research in molecular biology in China by Tien-hai (sic) Cheng and Roy H. Doi.

277 CHENG, Tien-hsi and DOI, Roy H.
Recent nucleic acid research in China. *Federation Proceedings*, 27(6), 1430-54, 1968. Also *Progress in nucleic acid research and molecular biology*, J.N. Davidson and W.E. Cohen (eds.). New York, Academic Press, Vol. 8, P. 335-58, 1968.
General support of experimental biology by the Communist regime. Centers of research and personnel; equipment and facilities. Discussion and evaluation of research activities published in the preceding 6 years, summaries of the investigators' findings and conclusions. Approximately 60% of papers reviewed meet Western standards.

278 *CHINA NEWS ANALYSIS*
The science of agriculture; genetics, No. 394, 27 Oct 1961.
Agricultural research in the post Great Leap period. Chinese geneticists abstain from philosophizing.

279 DOVER, Cedric
The use of biology in China. *United Asia*, 8(2), 134-6, 1956.
Observations based on the author's visit to six universities and three medical colleges. Research in entomology. Laboratory of Vertebrate Palaeontology. Anthropological research under the Institute of Experimental Biology.

280 ERRERA, M.
Molecular biology in China. *Nature*, 205(4973), 739-41, 1965.
Report of a visit to China by Belgian scientists. Examples of Chinese text books being prepared by members of the Biochemical Institute in Shanghai. Protein chemistry; the respiratory system; nucleic acids.

281 GALSTON, Arthur W.
No grades, no tests. *Yale Alumni Magazine*, 35(7), 8-11, 1972.
General observations of changes in the education system of China, observed by the author on his 1971 visit to China. Brief description of biology department at Chungshan University. '. . . there is essentially no basic research going on in China now . . .' but the Chinese 'recognise the ultimate necessity of returning to basic research when their physical conditions permit this to be done.'

282 LI, C.C.
Genetics and animal and plant breeding. Sciences in Communist China, Gould. *AAAS* No. 68, P. 297-321, 1961. Also *The China Quarterly*, No. 6, P. 144-152, Apr/Jun 1961.
Review of the debate among Chinese geneticists of the Lysenko and conventional schools. Since the Tsingtao Conference, the two schools have co-existed. Applied research may be a more important measure of China's scientific achievements than originality and quality of scientific work.

283 LIM, Robert K.S. and WANG, G.H.
Physiological sciences. Sciences in Communist China, Gould. *AAAS* No. 68, P. 323-62, 1961.

284 WADDINGTON, C.H.
Biology in China. *Arts and sciences in China*, 1(3), 2-5, 1963.
Observations on biological research made by the author during visit to China, Sept-Oct 1962, as a member of the Royal Society delegation.

## Botany

285 LI, Hui-lin
Botanical sciences. Sciences in Communist China, Gould. *AAAS* No. 68, P. 161-95, 1961.
Organization of botanical institutes. Review of research in taxonomy, mycology and microbiology; ecology and geobotany; morphology and cytology; plant physiology; economic botany. In general, there is now greater activity in botanical research and general studies in botanical sciences.

## Zoology

286 CHENG, Tien-hsi
Zoological sciences since 1949. Sciences in Communist China, Gould. *AAAS* No. 68, P. 197-226, 1961.
Entomology, ornithology, hydrobiology, parasitology. Contributions by Chinese zoologists are largely descriptive or observational reports, especially in taxonomy, morphology,

ecology and distribution. Some limited basic research of commendable quality by senior scientists trained in Europe or America.

## Biochemistry

287 MCELHENY, Victor K.
Total synthesis of insulin in Red China. *Science,* 153(3733), 281-3, 1966.
'The existence of the project (total synthesis of bovine insulin which possesses the full biological activity of natural insulin) indicates that China is making a modest but significant effort in fundamental biochemistry.' Review of the Chinese techniques.

288 VON HOFSTEN, Bengt
Kemi och politik in Folkrepubliken Kina (Chemistry and politics in the People's Republic of China). *Kemisk Tidskrift,* No. 11, P. 30-2, 1972.
General information on the political situation influencing the development of scientific research and higher education. Institutions of biochemistry and plant physiology in Shanghai were the best equipped of institutions visited. Much of applied research in microbiology and biochemistry is geared to the development of food production. The very large number of 'biotechnical' factories are a consequence of the emphasis on application of microbiological research.

See also the following entries:

Policy for Science. 131, 148, 149, 173, 189, 190, 210.
Policy Toward Science. 228, 249.
Scientific Communication. 374, 378, 393.
Medicine and Public Health. 693.

## Geology and Earth Sciences

289 CHAO, E.C.T.
Progress and outlook of geology. Sciences in Communist China, Gould. *AAAS* No. 68, P. 497-522, 1961.
Purposes: to survey the progress made in various branches of geology since 1950; to evaluate published work; to project future course of geology in China; to acquaint American geologists with activities in geology in Communist China.

290 *CHINA NEWS ANALYSIS*
Geography, No. 267, 6 Mar 1959.
Geography conference. The 1959 programme. The natural regions of China. George B. Cressey. The Shen Pao atlas.

291 HARLAND, W.B.
The organization of geology overseas: China. *Proceedings of the Geological Society of London,* No. 1633, P. 102-7, 23 Sep 1966.

292 HSIEH, Chiao-min
The status of geography in Communist China. *The Geographical Review*, 49(4), 535-51, 1959.
Influence of the Soviet pattern on geography in China; the Institute of Geography; geographical periodicals; trend of geographical studies--influence of the Soviet Union, denigration of Western geographers, geography treated as a physical rather than a social science; geographical education; two main fields of geography are physical and economic geography. Geography recognized as important in surveying environment and natural resources, planning development of irrigation, land use, soil erosion, transportation and hydrography.

293 OLDHAM, C.H. Geoffrey
Earth sciences in the People's Republic of China. *Proceedings of the Geologists' Association*, 78(1), 157-64, 1967.
Observations by the author on his 1965 visit to the Institute of Geophysics and Meteorology: seismology division, physics of the earth's interior, meteorological division, magnetics division; and to the Institute of Geology.

294 OLDHAM, C.H. Geoffrey
*Visits to Chinese institutes of earth science.* Institute of Current World Affairs, unpublished newsletter CHGO-48, 19 Aug 1965.
Statistical and technical information on Geophysical Research Institute of Academia Sinica and Institute of Geology.

295 WIENS, Herold J.
Development of geographical science, 1949-1960. Sciences in Communist China, Gould. *AAAS* No. 68, P. 411-81, 1961.
Nature of geographic work: meteorological and climatological development; hydrologic studies; oceanography, coastal geomorphology and mineralogy; desert phenomena and control of deserts; glaciology; limnology; scheme of natural geographic regions; economic geography; cartography.

See also the following entries:

Technology Policy:
Innovation and the Direction of Technological Change. 97.
Policy for Science. 125, 131, 149, 173, 200, 208.
Scientific Communication. 374, 378, 393.

## Meteorology

296 *CHINA NEWS ANALYSIS*
Meteorology, No. 363, 10 Mar 1961.
A brief history. Meteorological research. Meteorology and the communes.

297 RIGBY, Malcolm
Meteorology, hydrology, and oceanography, 1949-1960. Sciences in Communist China, Gould. *AAAS* No. 68, P. 523-614, 1961.

See also the following entries:

Policy for Science. 149, 208.
Research and Development:
  Geology and Earth Sciences. 293, 295.
  Geophysics. 298, 302, 303.
Scientific Communication. 374.

## Geophysics

298 JAPANESE SCIENTIFIC DELEGATION
(Natural Science) that visited Communist China. *Reports*. Tokyo, Japan-China Friendship Society, Sep 1966. Trans. *JPRS* 46226, 9 Aug 1968.
Status of Chinese seismological research reported; Chinese meteorological buildings visited; Kuo Mo-jo discusses present-day China.

299 OLDHAM, C.H. Geoffrey
*A Chinese conference on tectonics*. Institute of Current World Affairs, unpublished newsletter CHGO-9, 20 Aug 1962.
Translation of article in July 1962 *K'o-hsueh T'ung-pao* which gives signs of a change in the Chinese Government's attitude towards science. Two major trends are emerging: a closer study by the Chinese of scientific developments in Western countries, and more attention to basic research and greater academic freedon.

300 WILSON, J. Tuzo
Geophysical institutes of the USSR and of the People's Republic of China. *American Geophysical Union, Transactions*, 40(1), Mar 1959.
Part 2--China, P. 16-24. Observations made by the author on 1958 trip to China.

301 WILSON, J. Tuzo
Geophysics. Sciences in Communist China, Gould. *AAAS* No. 68, P. 483-96, 1961.
Observations based on visit to China.

302 WILSON, J. Tuzo
Mao's Almanac. *Saturday Review*, P. 60-4, 19 Feb 1972.
Observations of a Canadian geophysicist on his second visit to China, in autumn 1971, to cities and institutes first visited in 1958. Lines of research in geophysics in China: 'the Chinese have information that we lack in our efforts to predict earthquakes. This information has not been reported in the West.' The author lectured on plate tectonics; Chinese maps appear to support these ideas, but Chinese scientists 'grasped my Western ideas only slowly and vaguely.' Development of the observational techiques of Caltech professor Hugo Benioff in China. Priority on earthquake control during the Cultural Revolution: meteorology and prospecting now in separate institutes from geophysics;

Institute staff trebled in size, 1/3 of the staff being employed at the Institute, 2/3 in factories manufacturing geophysical instruments or attached to seismographic and geomagnetic stations in the countryside. Members of the revolutionary committees that ultimately govern research . . . I judge . . . to have been well chosen.'

303 WILSON, J. Tuzo
Red China's hidden capital of science. *Saturday Review of Literature*, Vol. 16, P. 47-56, 8 Nov 1958. Abridged in *Science Digest*, Vol. 45, P. 30-5, Fall 1959.
Account of the author's 1958 visit to China. Peking: Institute of Geophysics and Meteorology; The Peking Central Geophysical Observatory; The Institute of Geological Prospecting. Lanchow: local branch of the Academy of Sciences--the science library; The Lanchow Geophysical Observatory; the University of Lanchow.

See also the following entries:

Policy for Science. 173, 181, 208.
Research and Development:
  Geology and Earth Sciences. 294.
General Purpose Data Collection. 364

## 2 Industrial R & D

See the following entries:

Technology and Economic Growth. 7
Technology Policy:
  Innovation and the Direction of Technological Change. 69, 70, 72, 92, 97.
Policy for Science. 119, 121, 138, 153, 189, 191, 196, 202, 205, 209, 210.
Scientific Communication. 368.
Industrial Technology:
  Electronics Industry. 507, 509, 512, 513, 516, 518.
  Petroleum Industry. 534, 538, 539, 561, 562.
  Machinery Industry. 590.
  Basic Metals Industries. 599, 610.
  Mining Industry. 617, 626.
Engineering. 637.

## 3 Agricultural R & D

304 NASH, Ralph G. and CHENG, Tien-hsi
Research and development of food resources in Communist China. *BioScience*, Pt. 1, P. 643-56, Oct 1965. Pt. 2, P. 703-10, Nov 1965.
Organization of research: basic research by the Chinese Academy of

Sciences, applied research emphasized in Chinese Academy of Agricultural Sciences, research in agricultural colleges; planning of research; quantity and quality and utilization of manpower; dissemination of agricultural technology. Lines of research and organization of research in crop sciences, soil sciences, animal sciences.

305 OLDHAM, C.H. Geoffrey
*Visits to Chinese research institutes and scientific instrument factories*. Institute of Current World Affairs, unpublished newsletter CHGO-41, 11 Jan 1965.
Observations of the Chinese Academy of Agricultural Science, Kiangsu; Nanking Agricultural Mechanization Research Institute; Soil Science Research Institute.

306 PHILLIPS, Ralph W. and KUO Leslie T.C.
Agricultural science. Sciences in Communist China, Gould. *AAAS* No. 68, P. 227-96, 1961. Also in *The China Quarterly*, No. 6, P. 133-43, Apr/Jun 1961.
Agricultural policy; Soviet influence. Trends and accomplishments of agricultural science. Organization for agricultural improvements: agricultural policy making and implementation; organization of agricultural research; training of agricultural leaders. Although little that is new in the field of agricultural science has emerged under the Communist regime, major steps have been taken to train a corps of agricultural scientists and to build up research institutes and experiment stations. The Chinese Communists have made bedfellows of the old and the new, of folklore and the findings of the laboratory.

See also the following entries:

Technology Policy:
  Innovation and the Direction of Technological Change. 69, 97.
Policy for Science. 121, 138, 170, 176, 182, 196, 202.
Research and Development:
  Biology. 278.
  Military/Space Exploration and Travel/Nuclear Energy. 359.
Scientific Communication. 367, 374, 375, 378.
Extension of Scientific and Technical Information/Popularization of Science. 458.
Agricultural Technology. 466, 467, 470, 471, 476, 484, 488, 492, 495.

## 4 Medical R & D

See the following entries:

Technology Policy:
  Innovation and the Direction of Technological Change. 97.
Policy for Science. 148, 149, 167, 170, 196, 202.

Research and Development:
Military/Space Exploration and Travel/Nuclear Energy. 359.
Medicine and Public Health. 685, 686, 688, 690, 693, 700, 702, 703, 704, 707, 715, 716, 729, 731.

# 5 Military, Space Exploration and Travel, and Nuclear Energy R & D

307 AARKROG, A. and LIPPERT, J.
Comparison of relative radionuclide ratios in debris from the third and the fifth Chinese nuclear test explosions. *Nature*, 213(5080), 1001-2, 1967.

308 ABELSON, Phillip H.
The Chinese A-bomb. *Science*, 146(3644), 601, 1964.
Review of the technology underlying the first Chinese nuclear explosion.

309 *AVIATION WEEK & SPACE TECHNOLOGY*
Chinese push ICBM development, 87(7), 59, 1967.
Brief article on Chinese nuclear research program, speculation about development of missiles.

310 BARNETT, A. Doak
The inclusion of Communist China in an Arms-Control Program. *Daedalus*, P. 831-45, Fall 1960.
Review of the program of development of modern military technologies in China, atomic and nuclear research. Soviet scientific and technical assistance to China. Implications for international disarmament arrangements of China's military capacity.

311 BEATON, Leonard
The Chinese and nuclear weapons. *The Guardian*, 8 Feb 1962. Also in *Survival*, 4(3), 125-6, 1962. Also in *Military Review*, 42(11), 50-3, 1962.
Military and economic reasons suggest that China's resources be directed to basic industrial development; thus the decision to have a primitive bomb at the earliest possible date may have been reversed. China's nuclear program formally started in summer 1958 with opening of Soviet-assisted experimental reactor. Military and international political consequences of China's nuclear capability.

312 BEATON, Leonard
The Chinese bomb, The ISS View, Pt. I. *Survival*, 7(1), 2-4, 1965.
Technical and industrial background to China's first nuclear device. 'The balance of informed opinion definitely favours the use of gaseous diffusion by the Chinese.' The problem of developing a delivery system.

313 BEATON, Leonard and MADDOX, John
*The spread of nuclear weapons*. London, Chatto and Windus, for the Institute of Strategic Studies, 1962.
Chap. 7: China. The issue before Chinese policy-makers is whether

to strengthen China's military position and prestige with one or two nuclear explosions or to hold back until the nation has the industrial base to make nuclear weapons systems. Technical facilities in China: single reactor near Peking, supplied by the Soviet Union and dependent on the Soviet Union for fuel; speculation about a Chinese-designed uranium-fueled reactor. Problem of delivery systems. Chinese foreign policy and the bomb; Chinese military doctrine.

314 *BUSINESS WEEK*
China's shot tells a startling story, No. 1834, P. 144, 146, 148, 24 Oct 1964.
'The low yield of the first Chinese nuclear test had been anticipated, but not the advanced technology that the tests indicate the Chinese have.' Use of U235 'means that the Chinese have uranium enriching facilities such as gaseous diffusion plants or gas centrifuge plants. It also indicates that they have a far more complex production system than Western scientists and observers believed.' Subsequent Chinese tests will indicate whether the Chinese are yet capable of producing a 'predictability design curve' and of improving the yield-to-weight ratio. Speculation about research in progress: gas diffusion; fusion (thermonuclear) bomb.

315 CALDER, Nigel
Mao--1 in orbit. *New Statesman*, 79(2042), 612, 1970.
Political and military implications of the first satellite launched by the Chinese. Use of satellite for survey and communications not economically justifiable. Satellite launch indicates 'the Chinese have mustered a formidable array of talent and skill . . .' but diffusion of skills throughout the countryside has yet to be proved.

316 CASELLA, Alessandro
China's atomic tiger. *Far Eastern Economic Review*, 54(6), 353-4, 1966.
China's nuclear science and its related fields are apparently as good as that of the West. Limited industrial base and shortage of fissile material are constraints on the number of atomic weapons China possesses.

317 CHENG, Chu-yuan
Progress of nuclear weapons in Communist China. *Military Review*, 45(5), 9-15, 1965.
Review of the institutional framework within which nuclear and rocketry research is conducted; Soviet aid to the Chinese nuclear program; equipment and manpower. Impact of the Chinese bomb on international political situation, and impact on the research program of Sino-Soviet split. 'Communist China's economic capacity will inhibit the rapid development of nuclear weapons and a rocket program.'

318 DUCROCQ, Albert
Le Satellite Chinois. *Sciences et Avenir*, P. 448-55, Jun 1970.
To have placed 200 kilos in orbit, the Chinese must not only have

mastered space technology, but also have built up a metallurgical and an electronics industry.

319 EISENBERG, A. et al
Fresh fall-out in Israel from the second Chinese nuclear detonation. *Nature*, 210(5038), 833-4, 1966.
Scientific analysis of fall-out attributed to Chinese nuclear explosion of 14 May 1965.

320 EVANS, Gordon Heyd
China and the atom bomb. *Royal United Service Institution Journal*, Vol. 107, Pt. I, P. 30-4, Apr 1962; Pt. II, P. 130-4, May 1962.
The author attempts to reconstruct the history of China's atomic industry. 'China . . . enters the nuclear competition with a substantial ready-made advantage in the primary stages of uranium production.' Atomic reactors in China: 'Communist China is certainly able to build power units if it wishes.' Growth of China's electric power capacity may indicate building of gaseous diffusion installations. Nuclear weapon delivery system: 'Primitive delivery means.' 'At the present time (China) has a choice of three basic (modern delivery) systems: manned bombers, inter-continental ballistic missiles, or missile-carrying atomic submarines.' Advantages and disadvantages of each; general conclusions: 'First, China must have an intercontinental strike-force if she is to fulfil ambitious plans of conquest.' 'Second, Communist China must have a flexible long-range deterrent.

321 EVANS, Gordon Heyd
Communist China's A-Bomb program. *New Leader*, Vol. 44, P. 15-17, 18 Sep 1961.
Review of Chinese nuclear research facilities and achievements. 'Certain basic facts are clear': China has large deposits of uranium and thorium; it is likely that China has at least four large nuclear reactors; many experts believe that there have been several atomic explosions in China within the last six years; China appears to have developed ability to separate uranium isotopes; the Chinese are probably building ICBMs and IRBMs; China places great emphasis on science, especially the disciplines related to modern armaments--science education; technical facilities and manpower. 'While Moscow's scientific help could greatly speed up Peking's domestic weapons program, the Chinese can undoubtedly make a steady headway on their own.'

322 FEELY, H.W., BARRIENTOS, Celso, and KATZMAN, D.
Radioactive debris injected into the stratosphere by the Chinese nuclear weapon test of 9 May 1966. *Nature*, 212(5068), 1303-4, 1966.

323 FENG, Wen
Peiping's financial burden in developing nuclear weapons. *Chinese Communist Affairs*, 4(1), 45-6, 1967.
Estimated costs of Chinese nuclear research program.

324 FIX, Joseph E. III
China--the nuclear threat. *Air University Review*, P. 28-39, Mar/Apr 1966.

International reactions to explosion of China's first nuclear device. Review of the Chinese nuclear development program. Soviet assistance; institutional, material and manpower resources. The technology of the first tests. Speculation about future technological developments. Estimates of financial resources invested in the nuclear program. Speculation about China's future use of its nuclear military capabilities in international politics.

325 FRANK, Lewis A.
Nuclear weapons development in China. *Bulletin of the atomic scientists*, 22(1), 12-15, 1966.
Historical reconstruction of the 'evolution of the Chinese A-bomb': 1950-58, heavy reliance on the Soviet Union for financial, material and technical support; 1959-65, China almost entirely self-sufficient in nuclear weapons research, development, engineering, testing and production.

326 FUJII, Shoji
China's nuclear scientists. Trans. from Bungei Shunju, Tokyo. *Atlas*, 7(5), 297-9, 1964.
'According to 1958 statistics, there were only 340 associates and researchers in the nuclear energy department of Communist China's Academy of Sciences. The country itself had fewer than 500 nuclear scientists, including some 100 who were working in Soviet Russia.' 'Although the Soviet Union cut off aid to Communist China's nuclear production on 20 June 1959, it is safe to say that China's independent research on nuclear explosions had made great strides. Today the centers for this work reportedly include the Institute of Nuclear Energy Research of the Academy of Sciences, the University of Science and Technology in Peking and the Military Science Institute.' Brief profiles of China's 'leading atomic scientists': Ch'ien San-chiang; Ch'ien Hsueh-sen; Wang Kanchang; Li Ssu-Kuang; Wu Yu-hsun.

327 GAUVENET, Andre
Les Essais Nucleaires Chinois. *Revue de Defense Nationale*, Vol. 23, P. 1765-75, 23 Nov 1967. Trans. *JPRS* 43621, 7 Dec 1967.
Technology of the Chinese nuclear tests; scientific manpower; foreign aid. Conclusion: Soviet aid was indispensable to the rapid development of the Chinese program.

328 GUILLAIN, Robert
Ten years of secrecy. *Bulletin of the atomic scientists*, 21(2), 24-5, 1965.
Atomic research in China.

329 HALPERIN, Morton H.
*China and the bomb*. New York, Frederick A. Praeger, 1965.
Political and military implications of Chinese nuclear capability. Chap. 3--China's nuclear potential: Soviet nuclear assistance to China; Chinese development of nuclear weapons.

330 HALPERIN, Morton H.
China's nuclear strategy. *Diplomat*, Sep 1966. Also in *Survival*, 8(11), 350-3, 1966.
Soviet aid to Chinese nuclear weapons development program.

Technology used in Chinese nuclear weapons. China's image of the causes and likely evolution of a nuclear war. China's motives for developing nuclear weapons and its behavior as a nuclear power.

331 INGLIS, David R.
The Chinese bombshell. *Bulletin of the atomic scientists,* 21(2), 19-21, 1965.
The explosion of the Chinese atomic bomb was not unanticipated but the use of U235 in the bomb, acquired by gaseous diffusion, suggests that Chinese production capacity is greater than previously estimated.

332 JOHNSON, Chalmers
China's 'Manhattan Project'. *New York Times Magazine,* P. 23, 117-19, 25 Oct 1964.
History of nuclear research program in China.

333 JONAS, Anne M.
Atomic energy in Soviet Bloc Nations. *Bulletin of the atomic scientists,* 15(9), 379-83, 1959.
Development of atomic energy program in China. The reactor offered China is similar in design to the U.S. CP-3 reactor.

334 KISHIDA, Junnosuke
Chinese nuclear development. *Japan Quarterly,* Vol. 14, P. 143-50, Apr/Jun 1967. Also in *Survival,* 9(9), 298-304, 1967.
Deductions about China's nuclear development: 1) China is anxious to speed up the early development of practical nuclear weapons, using enriched uranium as fissionable material, or a hydrogen bomb; 2) the yield-weight ratio of nuclear weapons is being rapidly improved; 3) China has already deployed a nuclear weapon system. Aim of nuclear development: to create an effective nuclear deterrent against the United States and the Soviet Union; to make the military development of her nuclear missiles the propulsive force behind the development of her industrial technology. 'China's drive to develop her technology and modernize her industry is centered around her military technology.' Domestic and international military ramifications.

335 KRAMISH, Arnold
The Chinese People's Republic and the bomb. Santa Monica, *RAND* Corp., Publication P-1950, 23 Mar 1960.
An analysis of the technological potential of a research reactor (such as that purchased by the Chinese from the Soviet Union and commissioned in mid-1958) as the sole source of fissionable material for development of an independent atomic bomb capability by China. Conclusion: the Chinese heavy water research reactor will produce enough material for a 'primitive' Nagasaki-type weapon in approximately 4½ years, which is also sufficient for the necessary adjunct research, and for construction of plutonium separation and fabricating facilities. If the Chinese do not want to continue dependence on the Soviet Union for loadings of enriched uranium, they will use the research reactor for training and will construct reactors using heavy water and natural uranium.

This will require heavier capital expenditure, high levels of technical competence and of manufacturing ability, but they will result in a greater bomb production capability.

336 KRAMISH, Arnold
The great Chinese bomb puzzle--and a solution. *Fortune*, P. 157-9, 250, Jun 1966.
Attempt to reconstruct the technology which permitted rapid development of nuclear devices by the Chinese.

337 LAPP, Ralph E.
China's mushroom cloud casts a long shadow. *New York Times Magazine*, 14 Jul 1968.
History and organization of nuclear weapons and delivery system research in China; development of U235 rather than plutonium for H-bomb application; Chinese scientists trained in the West and the Soviet Union; training of 6000 lower-echelon scientists and technicians in the USSR. US strategy to counter threat of Chinese ICBMs.

338 LI, Hui-min
An analysis of Peiping's two recent nuclear tests. *Chinese Communist affairs*, 4(5), 1-3, 1967.

339 LIU, Chi-chuen
A study of the Chinese Communist nuclear program. *Issues and studies*, 1(2), 23-33, 1964.
Establishment of nuclear research organizations. Concrete measures in nuclear study.

340 MACIOTI, Manfredo
Mao Entra in Orbita. *Adesso*, P. 84-8, Jun 1970.
China's launch of a satellite, like its nuclear and chemical successes, reflects competition with the US and USSR.

341 MADDOX, John
The Chinese A-bomb--and who next? *New Scientist*, 24(414), 215-6, 1964.
Review of the history of the Chinese nuclear program. 'The chances are that the Chinese effort is comparable with the British effort almost a decade ago.' China's bomb is not yet a significant military threat, but may provoke other nations to undertake nuclear development programs.

342 *NEW SCIENTIST*
A Chinese bomb that went phut, 37(582), 231-2, 1968.

343 NIU, Sien-chong
Communist China and its nuclear weapons. *NATO's fifteen nations*, Apr/May 1970.
Review of the nuclear development program: Soviet assistance; the technology of China's atomic devices. 'The Chinese nuclear program from 1957 to the first test in 1964 cost approximately U.S.$2.5 billion; the annual cost in 1965 was estimated as U.S.$470 million . . .' It may be concluded that Communist China has made

significant progress towards the development of strategic nuclear force in spite of some delays and disruptions attributed to the Cultural Revolution.'

344 NIU, Sien-chong
Red China's nuclear might. *Ordnance Magazine,* Jan/Feb 1970.
Review of the nuclear development program by a 'civilian government adviser to the Republic of China' on national defense matters. Soviet assistance to Chinese nuclear development. Organization of nuclear research: the Institute of Atomic Energy; Chinese participation in research at Dubna--'Through this arrangement the Soviets could siphon off the efforts of the best Chinese and East European nuclear physicists.' Technology of the Chinese nuclear explosions. The Cultural Revolution.

345 NOVICK, Sheldon
The Chinese bomb. *Scientist and citizen,* Vol. 7, P. 7-11, Jan/Feb 1965.
Speculation about the technology underlying the first Chinese atomic bomb: constructed of U235, triggered by implosion of shaped charges of chemical explosive. 'The ability to produce uranium 235 has implications for civilian technology . . . . The test explosion of an atomic bomb may be only the side-product of a program of civilian nuclear power. Such a program would enable the Chinese to bypass many difficult and costly steps in the process of industrialization, obviating the need for, for instance, the extensive mining and transportation facilities required by conventional electrical power.'

346 *NUCLEONICS*
Red China claims to be firmly in atomic age, 17(1), 21-2, 1959.
Description of China's first reactor and cyclotron. Laboratory facilities, charts and uranium ore specimens on display at Peiping national industrial and communications exhibition. Separation by Chinese scientists of eleven spectroscopically pure rare earths from Chinese minerals.

347 PAO, Chin-an
Peiping's capacity for nuclear weaponry. *Chinese Communist affairs,* 5(2), 1968.

348 PERSSON, G.
Fractionation phenomena in debris from the Chinese nuclear explosion in May 1965. *Nature,* 209(5029), 1193-5, 1969.
Analysis of debris from second Chinese nuclear device explosion found in some high-altitude air-samples collected over Sweden.

349 PERSSON, G.
Observations on debris from the first Chinese nuclear test. *Nature,* 209(5209), 1228-9, 1969.

350 PERSSON, G. and SISEFSKY, J.
Debris from the sixth Chinese nuclear test. *Nature,* 223(5202), 173-5, 1969.
Analysis of debris collected over Sweden.

351 RIFKIN, Susan Beth
The development and use of nuclear energy in the People's Republic of China. *IR&T Nuclear Journal*, 1(4&5), 1969.
By developing atomic energy, China has gained a technological base for a new source of energy for civilian and military purposes; created a weapons program that has commanded world attention; and proved that China's approach can be used to bring a non-industrial country to the threshold of a modern industrial and military capability in a short time. Imports of training and technical aid were the foundation of the Chinese nuclear program; Soviet assistance relieved China of some of the financial costs of nuclear research; Soviet aid was not given for development of nuclear weapons, but influence was exerted to direct Chinese nuclear program towards peaceful applications, although Soviet aid is the basis of the nuclear weapons program. I--The foundations of nuclear research. II--Soviet assistance and Chinese nuclear weapons development.

352 *SCIENCE JOURNAL*
Chinese nuclear test--her second H-bomb? 3(8), 15-17, 1967.
Review of the technological progress of China's nuclear program, based on evidence obtained from fall-out from Chinese nuclear explosions.

353 *SCIENCE JOURNAL*
Technology behind China's second bomb, 1(5), 13, 1965.
Where Chinese nuclear engineering has deviated from the set pattern is in simply opting for enriched uranium from the start.

354 *SCIENCE NEWS*
China joins the space age, 97(18), 427-8, 1970.
Data on the first Chinese satellite indicate a medium-range ballistic missile capability; the science and technology required for ICBMs is more sophisticated and, although the theoretical knowledge may be available to the Chinese, shortage of engineering and logistical know-how and disruption during the Cultural Revolution may be the cause of the lag in Chinese ICBM development. International political and military repercussions of China's missile capabilities.

355 SISEFSKY, J.
Debris particles from the second Chinese nuclear bomb test. *Nature*, 210(5041), 1143-4, 1966.

356 THEIN, Myint and KURODA, P.K.
Global circulation of radiocerium isotopes from the 14 May 1965 nuclear explosion. *Journal of geophysical research*, 72(6), 1673-80, 1967.
Technical analysis of rainfall over Fayetteville, Arkansas indicates that the radiocerium isotopes produced travelled eastward around the world at least twice.

357 U.S. DEPT. OF STATE
Bureau of Intelligence and Research. *Nuclear research and technology in Communist China.* External Research Paper 139, Apr 1964.

A list of monographs and periodicals and newspaper articles in the English language on atomic energy research and technology in Communist China.

358 WAKAMATSU, Jugo
Powerful new weapons are mass-produced to China's own designs in 'Science and technology in modern China'. *The Asahi Asia Review*, 3(1), Spring 1972.

359 WANG, Chi
Nuclear research in Mainland China. *Nuclear news*, 10(5), 16-20, 1967.
Initial development of nuclear research facilities with Soviet aid. Organization of research facilities: Institute of Atomic Energy of the CAS, divided into two atomic research centers. Facilities of these centers, as described by Australian and Japanese visitors. Nuclear research in other branches of the CAS; Committee of Atomic Energy and Committee of Isotope Applications in the CAS. Application of atomic energy in agricultural sciences and medicine studied by Research Laboratory of Atomic Energy Applications of the Agricultural Sciences and by the Academy of Medical Sciences. Shortage of qualified scientific manpower; advanced training in Academy institutes and leading universities, and at Dubna. All cooperation with Soviet Union in nuclear research and training has been terminated. Nuclear research publications. Conclusion: 'nuclear research in Mainland China has passed the embryonic stage and more repid advances may now be anticipated.'

360 WANG, Chi
*Nuclear science in Mainland China, a selected bibliography*. Washington, U.S. Government Printing Office, 1968.
Pt. I--Items in Chinese by Chinese scientists and engineers; results of original research; some review and 'current awareness' articles. Pt. II--Items from scientific journals, leading periodicals, and news magazines published chiefly in the U.S., Japan; also English translations of Chinese materials and English-language publications in China. Review articles relating to the development and potential of China's nuclear program.

361 WILSON, Dick
China's nuclear effort. *Far Eastern Economic Review*, 49(8), 328-9, 1965.

362 WOLFSTONE, Daniel
China's research reactor. *Far Eastern Economic Review*, 27(9), 315-17, 1959.
10,000 KW research atomic reactor constructed with Russian aid in Peking. Chinese claim to have produced 33 kinds of radioactive isotopes. Nankai University, Tientsin, reported to have built experimental homogeneous reactor. Tsinghua scientists reported to have trial-produced induction electron accelerator.

See also the following entries:

Technology Policy:
  Innovation and the Direction of Technological Change. 92, 97.
Policy for Science. 116, 119, 123, 124, 129, 138, 141, 153, 156,
  157, 159, 160, 166, 172, 175, 180, 182, 203, 212.
Policy Toward Science. 243, 246.
Higher Education and Technical Training. 414.
Industrial Technology:
  Electronics Industry. 516.

## B General Purpose Data Collection

363 *CHINA NEWS ANALYSIS*
A scientific survey of land utilization, No. 562, 30 Apr 1965.
Review of article by four Chinese scientists on new division of China according to morphological features, average temperature, number of frost-free days and typical plants.

364 EPINAT'EVA, A.M. and KOSMINSKAYA, I.P.
Seismic prospecting in China. *Izv. Geophys. Ser.* P. 1673-83, 1959.
Review of seismic prospecting operations in China, focusing on methodological problems, as observed by authors on a visit to China in 1958 under auspices of CAS Geophysical Institute.

See also the following entries:

Policy for Science. 171, 210.
Research and Development:
  Geology and Earth Sciences. 292.
  Military/Space Exploration and Travel/Nuclear Energy. 315.
Industrial Technology:
  Petroleum Industry. 533, 536, 538, 539, 540, 544, 551, 561,
  562.
  Mining Industry. 616.
Natural Resources. 649, 651, 652.
Environmental Control. 756.

## C Scientific Communication

365 *CHEMICAL & ENGINEERING NEWS*
China: few scientific contacts, P. 6, 8 Nov 1971.
Scientific contacts between China and the U.S. are near zero. *Chemical Abstracts* once received 112 Chinese journals, but no journals have been received since 1966. In 1964, 240 Chinese subscriptions to journals of the American Chemical Society, but most were cancelled in 1966, and complimentary exchange subscriptions ceased in 1968. Last two ACS members in China resigned from the Society in 1967.

366 *CHEMICAL & ENGINEERING NEWS*
Chinese scientists in midst of U.S. tour, P. 21-2, 4 Dec 1972.
Report of the visit of seven Chinese scientists to the U.S.; brief biographies. At least 70 American scientists, engineers and physicians have visited China. 'Scientific exchanges are . . . seen by both the U.S. and Chinese governments as an important area for initiation of new relations between the two countries.'

367 CHENG, Tien-hsi
The Entomological Society of (Communist) China. *Bulletin of the Entomological Society of America*, 9(4), 266-7, 1963.
Organization and activities of the Entomological Society of China. The Plant Protection Society of China.

368 *CHINA NEWS ANALYSIS*
Science and research, Summer 1964. No. 536, 9 Oct 1964.
The Peking Science Symposium including list of Chinese contributions. Article in *JMJP* 20 Aug 1964 by Ch'ien San-ch'iang: 'Scientific experiments on a large scale'. *JMJP* editorial 5 Sep 1964, 'Look further ahead', on industrial research. Article by Pai Chieh-fu, deputy head of Chemistry and Physics Research Institute, CAS, in *JMJP* 18 Apr 1964, on application of research to production.

369 CHINA RESEARCH INSTITUTE (Tokyo)
The 1964 Peking Science Symposium 1. *China Research Monthly Bulletin*, No. 202, Dec 1964.
Ando Hikotaro, results of the Peking Science Symposium. Sakata Shoichi, head of the Japanese delegation (Li Szu-kuang, The Scientific Democratic Spirit at the Peking Science Symposium). Table of notes related to the Peking symposium; Table of topics of papers at the Peking symposium; Appendix: Organizational structure of the Chinese Academy of Sciences.

370 DIRECTORATE FOR SCIENTIFIC AFFAIRS
*Chinese scientific and technical literature*. DAS/CSI/66.198, 23 Jun 1966. Revised list, DAS/CSI/67.54, 4 Jul 1967 (circulation restricted) OECD, Paris.
List of Chinese scientific and technical journals regularly received in the West.

371 DIRECTORATE FOR SCIENTIFIC AFFAIRS
*Study on the utilization of scientific and technical literature from China*, DAS/RS/65.43, 19 Feb 1965 (circulation restricted) OECD, Paris.
'The Committee for Scientific Research . . . decided to undertake surveys to find out how far publications in Asian languages are available in member countries, what use is made of them, and what facilities exist for translation . . .' Analysis of holdings of Chinese scientific and technical literature by country.

372 *FAR EAST TRADE & DEVELOPMENT*
China-U.K. understanding, 27(10), 412-13, 1972.
Report of visit by a team of Chinese scientists to the UK, 7-20 October, 1972. List of participating scientists, description of

their interests and itinerary of visits. 'Primary purpose of the visit, according to the leader Professor Pei Shih-chang was to obtain a general picture of the leading areas of science in the UK, and to make contacts with people in those areas with whom Chinese scientists might liaise in future.'

373 KLOCHKO, Mikhail Antonovich
*Soviet scientist in China.* London, Hollis, 1964. Also Education and Communism in China, Stewart E. Fraser (ed.), P. 465-94.
Observations made by a top-ranking Soviet scientist while in China in 1958 and 1960 to help in development of research programs and to teach.

373a OKA, Takashi
Science in China: the other face. *Current scene,* 3(2), 1 Sep 1964.
Review of *Soviet scientist in Red China.*

373b TRETIAK, Daniel
Russian expert in China. *Far Eastern Economic Review,* 50(5), 257-60, 1965.
Review of M. Klochko, *Soviet scientist in Red China.*

374 KUO, Leslie T.C. and SCHROEDER, Peter B.
*Communist Chinese monographs in the U.S.D.A. Library.* Washington, U.S. Dept. of Agriculture Library, Jun 1961.
Provides a view of the variety of publications prepared in agricultural and related fields, including materials for all educational levels. Subject headings: agriculture; plant science; agricultural methods; forestry and forest products; animal science; fisheries; apiculture; sericulture; food and nutrition; processing of agricultural products; biology and chemistry; nation's economy; geology, geography, meteorology and pedology.

375 KUO, Leslie T.C. and SCHROEDER, Peter B.
*Communist Chinese periodicals in the agricultural sciences.* Washington, U.S. Dept. of Agriculture Library, Dec 1960. Rev. May 1963. Library List No. 70.
Revised edition: 132 titles arranged alphabetically by title in Wade-Giles romanization; list of publishers and subject index.

376 LEAR, John
Dispassionate scientific look at China. *Saturday Review,* Vol. 41, P. 45, 8 Nov 1958.
Science transcends international politics; the Communist and Nationalist Chinese could meet at the International Council of Scientific Unions.

377 LEAR, John
Global pollution and the U.N.; Pt. 4, The Chinese influence. *Saturday Review,* P. 41, 7 Aug 1971.
Speculation about possible participation by China in first United Nations conference on the human environment, June 1972.

378 LEE, Amy C. and DJU CHANG, D.C.
*A bibliography of translations from Mainland Chinese periodicals*

*in chemistry, general science and technology*, Published by U.S. Joint Publications Research Service, 1957-1966. Washington, National Academy of Sciences, Committee on Scholarly Communication with Mainland China, 1968.
Scope and purpose: to provide bibliographic access to the body of unclassified and undefined *JPRS* science and technology translations from Mainland Chinese sources; to provide bibliographic access to all chemical titles translated from Mainland Chinese sources. I--Periodicals in the specific sciences. II--Multidisciplinary journals and magazines. III--Periodicals in the social sciences and in the humanities. IV--Newspapers. V--Books and proceedings.

379 MACDOUGALL, Colina
The advancement of learning. *Far Eastern Economic Review*, 38(5), 270, 1962.
Report on visit to Royal Society delegation to China, 1962.

380 MASSACHUSETTS INSTITUTE OF TECHNOLOGY
*Current holdings of Communist Chinese journals in the M.I.T. libraries.* Cambridge, Mass., 1960.
136 titles of Communist Chinese scientific and technical journals in the M.I.T. libraries, arranged by science subject group.

381 MASSACHUSETTS INSTITUTE OF TECHNOLOGY LIBRARIES
*KWIC index to the science abstracts of China.* Cambridge, Mass., Dec 1960.
This computer generated index is a key word and author index to all issues of the Communist Chinese publication, *Science abstracts of China*, presently known to be available in the U.S. It provides a ready reference to 3300 current Communist scientific and technical papers. Pt. I--Alphabetic listing of key words-in-context in the titles of papers abstracted. Pt. II--Listing, by author, of title, source and full bibliographic data. Pt. III--Index of authors of papers included.

382 NATIONAL LENDING LIBRARY FOR SCIENCE AND TECHNOLOGY.
*List of scientific and technical periodicals received from China.* Boston Spa, 1964.
Holdings of scientific and technical periodicals received from China; Key-word index; Wade-Giles index.

383 NATIONAL LIBRARY OF MEDICINE
*Current holdings of Mainland Chinese journals.* Bethesda, Maryland, 1965.
Pt. I--Alphabetical list of journals. Pt. II--List of publishers. Pt. III--Variant title index.

384 *NATURE*
Does China exist, 215(5096), 3, 1967.
Editorial urging greater international scientific contacts by China.

385 *NATURE*
Old World: welcome visitors, 239(5370), 245, 1972. New World:

visitors from China, 239(5371), 302, 1972. New World: more cracks in the ice, 240(5375), 6, 1972.
Visit by a multi-disciplinary scientific delegation from China to Britain. Visit by ten Chinese medical doctors to the U.S. Background on arrangements for the November 1972 visit of seven Chinese scientists to the U.S.

386 NUNN, G. Raymond
*Publishing in Mainland China,* M.I.T. Report No. 4. Cambridge, The M.I.T. Press, 1966.
P. 29-30 Science Publishing, including tabulated statistics: development of the Science Press, 1954-1963.

387 OLDHAM, C.H. Geoffrey
*The Peking Science Symposium.* Institute of Current World Affairs, unpublished newsletter CHGO-33, 20 Sep 1964; Pt. II, CHGO-42, 11 Jan 1965.
Report of Peking Science Symposium, 21-31 Aug 1964, based on official Chinese sources. Table of papers presented. Comment on the Symposium as a scientific conference where new work is presented and critically discussed; as a conference on the application of science and technology to development; as a Chinese propaganda device; as a reflection of international political tensions; as a rally of scientists from developing countries.

388 OLDHAM, C.H. Geoffrey
*Stirrings of British interest in Chinese science.* Institute of Current World Affairs, unpublished newsletter CHGO-3, 25 Apr 1961.
Four categories of interest in Britain in Chinese science: individual scientists connected with academic institutions; scientific societies and professional associations; universities and technical colleges teaching Chinese to scientists; government groups and industrial organizations. Interest of Mr. Robert Sewell of United Steel Company in Chinese science.

389 'D.S.'
Year of the dove? *Science,* 172(3982), 457, 1971.
International contacts with Chinese scientists in the wake of new Chinese diplomatic policies in 1971: Chinese participation at National Accelerator Laboratory (Batavia, Ill.); National Academy of Science/Committee on Scholarly Communication with Mainland China; communication with individuals in China; Chinese participation at Pugwash. 'But China herself may not yet be ready for scientific relations abroad.' Leo Orleans: Chinese scientists transmit research information via the centralized bureaucracy, but publication will resume gradually.

390 SEWELL, R.
The problem of Chinese technical literature. *The British steelmaker,* P. 92-5, Mar 1960.
The bulk of the useful technical and scientific literature is being produced in Chinese only, and may present technical information services with a problem of translation, abstracting and dissemination in five to ten years.

391 SHIH, Bernadette P.N. and SNYDER, Richard L.
*International union list of Communist Chinese serials; scientific, technical and medical, with selected social science titles*. Cambridge, M.I.T., 1963.
Holdings of over 500 recent Communist Chinese scientific, technical and medical serials in 28 libraries of the U.S., Canada, Great Britain, Japan and Hong Kong plus a list of 100 social science titles.

392 TSAO, Chia Kuei
*Bibliography of mathematics published in Communist China during the period 1949-1960*. Providence, American Mathematical Society, 1961.
Chinese periodicals containing mathematics bibliography (alphabetical, by author). The number of mathematical articles published yearly has been increasing; there are indications that the Chinese Mathematical Society is placing more emphasis on the application of mathematics.

393 U.S. DEPT. OF COMMERCE, OFFICE OF TECHNICAL SERVICES
Chinese Mainland science and technology. *OTS selective bibliography* SB-442, Supplement 1. Washington, n.d.
Agriculture; biological and behavioral sciences; chemistry; earth sciences; engineering; machinery; metallurgy; physics and mathematics; research methods, techniques and equipment; bibliographies.

394 U.S. LIBRARY OF CONGRESS
*Chinese scientific and technical serial publications in the collections of the Library of Congress*. Washington, Science Division, 1955. Rev. 1961.
The bibliography is presumably complete for all serials containing reports of research and other significant scientific information. Contents: science, general; pure science; technology, general; technology; agriculture; medicine; abstracting, indexing and bibliographical services.

395 VAN DEN BERGAN DE GEER, C.A.
*China: guide of books, periodicals and records in the library of the Delft Technological University*. Library of the Delft Technological University, 1965.
Includes scientific and technological dictionaries; Chinese periodicals; books on science and technology acquired 1960-65.

396 WARNER, Charles
Developing science. *Far Eastern Economic Review*, 46(2), 68-70, 1964.
Peking Science Symposium, 1964.

397 WINCHESTER, John
Importance of Chinese for scientific communication. *Science*, 131(3412), 1561-2, 1960.
Letter to the editor on the desirability of training Western scientists in the Chinese language. 'In view of the rapidly

increasing importance of Chinese scientific literature it is desirable that some scientifically trained persons now begin learning to read scientific Chinese. Only when knowledge of the language is widespread can Chinese scientific progress be evaluated accurately.'

398 WOLFLE, Dael
Chinese embargo. *Science*, 134(3475), 303, 1961.
Chinese embargo on export of scientific journals prevents knowledge of scientific research in China and prevents Chinese scientists from keeping up with Western literature, since exchange arrangements are cut off.

399 YAMANAKA, Akio
On Sino-Japanese academic (physics) communication. *Japan-China*, 2(4), 1, 1972.
Records of a symposium on Sino-Japanese academic exchanges.

See also the following entries:

Policy for Science. 119, 125, 127, 131, 133, 139, 144, 154, 171, 175, 176, 184, 190, 191, 194, 197, 208, 210, 212.
Policy toward Science. 222, 229.
Research and Development:
  Mathematics. 258.
  Physics. 264.
  Astronomy. 270.
  Electronics. 271.
  Geology and Earth Sciences. 292.
  Geophysics. 299.
  Nuclear Energy. 359.
Higher Education and Technical Training. 452.
Industrial Technology:
  Electronics Industry. 516, 517.
Medicine and Public Health. 685, 690, 707, 711.
Population Control. 745.

## D Higher Education and Technical Training

400 ABE, Munemitsu
Spare-time education in Communist China. *The China Quarterly*, No. 8, 149-59, Oct/Dec 1961. Also *Education and Communism in China*, Stewart E. Fraser (ed.). London, Pall Mall Press, P. 239-53, 1971.
The spare-time school aims to provide education for cadres, workers, and peasants; includes 'spare-time universities', literacy classes, elementary schools for former illiterates, refresher courses for teachers, semi-professional education. Scientific and technical education in spare-time schools.

401 ABELSON, P.H.
Mainland China: an emerging power. *Science*, 157(3787), 373, 1967.
Editorial on the 'intellectual potential of the Chinese': education and training of professional manpower.

402 ALLEY, P.J.
Some engineering universities in China. *Eastern Horizon*, 5(8), 12-14, 1966.
Observations of polytechnical institutes made by the author on 1957-58 and 1966 visits to China.

403 *ARTS AND SCIENCES IN CHINA*
A science-minded nation, 1(1), 3-5, 1963.
Science education in China; cost to research in emphasizing education. Popularization of science.

404 BASTID, Marianne
Economic necessity and political ideas in educational reform during the Cultural Revolution. *The China Quarterly*, No. 42, P. 16-45, Apr/Jun 1970.
Reforms aim at removing narrow specialization in education. Science graduates outnumber employment places.

405 BERI, G.C.
The development of education and professional manpower in China and India: a comparative study. *China Report*, 5(4), 1-8, Pt. I, 1969. No. 5, Pt. II, Aug/Sep 1969.
'This paper attempts to study some important aspects of the development of education and professional manpower in China and India. The study is divided into two parts. The first deals with the educational patterns as well as the growth of different levels and types of education in the two countries. A comparative study of the growth of selected categories of professional manpower is attempted in the second.'

406 BERNAL, J.D.
Science and technology in China. *Universities Quarterly*, 11(1), 64-75, 1956.
Impressions of the Chinese education system based on 1954 visit to China. 'The drive for rapid increase of trained manpower and the emphasis on utility for national development are the major factors determining the present character of Chinese universities and technical colleges.' After the reorganization of education in 1952, the general pattern of overall organization followed has been the Soviet one, but because many Chinese science teachers were trained on the British pattern, Chinese higher education resembles the British more than the Soviet. All universities emphasize science subjects over the humanities.

407 CARTIER, Michel
Planification de l'enseignement et formation professionnelle en Chine Continentale (Planning of education and professional training in Mainland China). *Tiers-Monde*, Vol. 6, P. 511-30, Apr/Jun 1965.
'The object of this study is to trace the main lines of

educational policy of the government of mainland China, with particular emphasis on the demographic and economic context of the measures adopted in planning.' Three objectives of Chinese policy: 1) elevation of the general education level; 2) formation of administrative cadres and technicians necessary for economic development; 3) development of a favorable attitude toward socialism. Pt. I--Education to the end of the First FYP. Pt. II--Reforms of 1958 and consequences.

408 CH'EN Feng-chi
Scientific and technical education in China. *Ajiya Kenkyu*, No. 387, P. 6-15, 10 Sep 1963. Trans. *JPRS* 22237, 11 Dec 1963. Survey of policies for scientific manpower and training in China from 1949 to date. Based largely on JMJP articles. June 1949: survey of S & T manpower by Preparatory Committee for the first conference of scientific and technological personnel of China. November 1959 (sic; 1949?): activities of CAS in connection with survey of scientific manpower. Recruitment of research trainees, conditions of employment and training and distribution among institutes of the CAS, 1950-7. 1958: The Chinese Institute of Science and Technology, supervised by CAS and Ministry of Education. System of science prizes and list of recipients, Nov 1955-March 1956. Increase in S & T manpower.

409 CHEN, Theodore Hsi-en
Education in Communist China. *Communist China, 1949-1969: a twenty-year appraisal*, Frank N. Trager and William Henderson (eds.) Chap. 8. New York, New York University Press, 1970. P. 184-5: Science and technology. 'The study of theoretical science . . . does not receive as much attention as engineering and the applied sciences. When the Communists talk about science, they actually mean technology.' Estimates of enrollment, manpower based on Chu-yuan Cheng, *Scientific and engineering manpower in Communist China* (see entry No. 412).

410 CHENG, Chu-yuan
Peking's minds of tomorrow: problems in developing scientific and technological talent in China. *Current scene*, 4(6), 1966. Also *Education and Communism in China*, Stewart E. Fraser (ed.). London, Pall Mall Press, P. 389-406, 1971.

411 CHENG, Chu-yuan
Scientific and engineering manpower in Communist China. *An economic profile of Mainland China*, U.S. Congress, Joint Economic Committee. New York, Praeger, P. 519-47, 1968.

412 CHENG, Chu-yuan
*Scientific and engineering manpower in Communist China 1949-1963*. National Science Foundation No. NSF 65-14, 1965. Also *Education and Communism in China*, Stewart E. Fraser (ed.). London, Pall Mall Press, P. 495-538, 1971.
The focal point of this study is scientific and engineering manpower, within the framework of general domestic and foreign policies. Original and official sources explored: 30 scientific and technological periodicals, 20 Party newspapers; English,

Russian, and Japanese, language materials. Conclusion: substantial strides in the fields of science and technology made during 1949-1963; steady progress during first FYP, setbacks during 1960-63. Soviet assistance and rapid expansion of training programs augmented the ranks of professional manpower. 2--Policy and planning for scientific development. 3--Training programs for scientists and engineers. 4--Quantity and quality in the training of scientists and engineers. 5--Employment of scientists and engineers. 6--Utilization of scientists and engineers. 7--Role of the Soviet Union in developing scientific and technical manpower in Communist China. 8--Role of Western-trained scientists and engineers. Appendix: I--Official documents. II--Board members of the department committees of the CAS, 1963. III--Officers of departments and research institutes of the CAS, Chinese Academy of Agricultural Sciences, and Chinese Academy of Medical Sciences, 1964. IV--Biographic data of 1200 prominent scientists and engineers.

413 *CHINA NEWS ANALYSIS*
Training in medicine, No. 577, 20 Aug 1965.
Medical training, 1961-1965; Chinese medicine.

414 *CHINA REPORT*
Higher education in science and technology in China, 3(1), 30-5, 1967.
Effect of Sino-Soviet dispute on Twelve Year Plan for development of science and technology: continued use of Western science and technology knowledge; Chinese students in Western Europe in the 1960s; impressive numbers trained but senior researchers are in a minority; accent on technical education; spare-time education; science for defense. The Academy of Science.

415 *CURRENT SCENE*
China's 'reformed' universities: the first year, 9(6), 9-10, 1971. 'According to (a report by New Zealand writer Rewi Alley published in the Hong Kong Communist newspaper, *Ta Kung Pao*), Peita is now organized into 18 departments in literature and the humanities and eight in the sciences . . . . The science departments include: physics, radio, sanitation, geology, biology, mathematics and mechanics.'

416 DEDIJER, S.
The sixth column. *Science,* 153(3738), 852-4, 1966.
Review of *Scientific and engineering manpower in Communist China, 1949-1963* (C.Y. Cheng, entry No. 412).

417 DIMOND, E. Grey
Medical education in China. In 'Science and medicine in the People's Republic of China'. *Asia,* No. 26, P. 60-73, Summer 1972. 'Lateral mobility' in the Chinese system of medical education: training in one medical profession (e.g. nursing) is regarded as preparation for another (e.g. medical practice). Medical training lasts three years: first year--basic science; second year--outpatient or outreach work; third year--hospital training, Continuing education after graduation. Mass health education.

418 ELLIOT, Denis
Spare-time studies. *Far Eastern Economic Review,* 55(12), 505-6, 1964.
Part-time universities and academies produce theoreticians, technical experts, advisers and designers who are red and expert.

419 EMERSON, John Philip
Employment in Mainland China: problems and prospects. *An economic profile of Mainland China,* U.S. Congress: Joint Economic Committee, P. 403-69.
Section on structure of employment: level of skill; engineering and technical personnel in industry and capital construction; level of educational attainment of engineers and technicians; levels of professional attainment in medicine and public health.

420 FRASER, Stewart
*Chinese Communist education: records of the first decade.* New York, John Wiley & Sons, Inc., 1965.
'The compilation of this series of select documents on Chinese Communist education has been undertaken so that some of the more useful statements would be set forth in a unified and readily available form for those interested in both the comparative and political aspects of contemporary Chinese education . . . . The documents principally encompass the period from 1950 to 1960, especially the 'period of so-called "intellectual freedom" from 1955 to 1957.' Contents: Stewart Fraser, Education, indoctrination, and ideology in Communist China, P. 3-69; Speeches, articles and documents on education from Communist China, P. 71-407; Bibliography of primary and secondary materials on education, in 17 sections. Section XV: Educational aspects of science, technology and economic development.

421 FRASER, Stewart E. (ed.)
*Education and Communism in China.* London, Pall Mall Press, 1971.
Leo A. Orleans, Quality of education, P. 81-103. Munemitsu Abe, Spare-time education in Communist China: a general survey, P. 239-53. Paul Harper, Problems of industrial spare-time schools, P. 255-74. Leo A. Orleans, Education and scientific manpower, P. 363-87. Cheng Chu-yuan, Problems in developing scientific and technological talent in China, P. 389-406. Mikhail A. Klochko, Science and scientists in Peking, P. 465-94. Cheng Chu-yuan, Role of the Soviet Union in developing scientific and technical manpower in Communist China, P. 495-538.

422 GALSTON, Arthur W.
The University in China. *BioScience,* 22(4), 217-20, 1972.
Views derived from the author's visit in May 1971 to China. Organization of the biology program at Chungsan (sic) University, Canton. There is essentially no basic research in China at present, although the Chinese recognize the necessity of carrying out basic research when 'their physical conditions permit this to be done.' The author's encounters with individual Chinese scientists whom he had known previously, and their experiences in the Cultural Revolution.

423 HARPER, Paul
Closing the education gap: problems of industrial spare-time schools. *Current Scene*, 3(15), 1-13, 1965. Also *Education and Communism in China*, Stewart E. Fraser (ed.). London, Pall Mall Press, P. 255-74, 1971.
'In the West the industrial revolution generated an evolution in education . . . in China, the regime has chosen to revolutionize education to anticipate industry's demands.' The industry-operated spare-time school system has become one of the most important networks for long-range development in China. Party and trade union administration of spare-time schools; political and cultural/technical education. History of spare-time education 1950-55, Great Leap Forward, retrenchment in 1960s. Jointly-run schools; part-work, party-study schools.

424 HARPER, Paul
*Spare-time education for workers in Communist China*, No. 30. Washington, U.S. Dept. of Health, Education and Welfare, Office of Education Bulletin OE-14102, 1964.
'The Chinese Communists, recognizing that raising the technical levels of the working force is a necessary adjunct of industrialization, have long given priority to workers' education in their plans for national development.' This review of workers' education will indicate the allocation of priorities in time spent and the content of workers' education in periods of crisis and of relative calm. It will also explore the problems and accomplishments of a dozen years of industry-centered learning, as well as the organization, conduct and control of the schools.'

425 HIKOTARO, Ando
Record of a visit to Peking University. *Ajia Keizai Jumpo*, No. 556, P. 6-10, Nov 1963.
Visit to Peking University, 28 Aug 1963. Interview with Professor Chou Pei-yuan, physicist and vice-president of the University.

426 HSU, Immanuel C.Y.
The impact of industrialization on higher education in Communist China. *Manpower and education: country studies in economic development*, Frederick Harbison and Charles A Myers (eds.). New York, McGraw-Hill, P. 202-31, 1965. Also The re-organization of higher education in Communist China, 1949-1961. *The China Quarterly*, No. 19, P. 128-60, 1964.
Anticipation of economic development and conscious planning for industrialization decisively change the nature and content of higher education. Technical education precedes industrialization. Universities and technical institutes train personnel to create industrialization.

427 HU, C.T.
Communist education: theory and practice. *The China Quarterly*, No. 10, P. 84-97, 1962.
'Within the ideological framework, scientific education means specifically the development of modern science and technology to expedite the process of industrialization and national development.'

428 HU, Chang-tu
*Chinese education under Communism.* New York, Columbia University Press, 1962.
Tradition and change in Chinese education. Combination of education and productive labor; emphasis on scientism will eventually create a generation of technically qualified workers who will transform China into an industrial power.

429 IKLE, F.C.
The growth of China's scientific and technical manpower. *RAND memorandum RM-1893,* 24 Apr 1957.
'This study is concerned primarily with the future capability of Communist China's educational system and with the expansion of its scientifically trained manpower . . . .' Shortage of scientists and technically trained personnel is probably a less important constraint on industrialization than shortage of capital. Contents include: elementary education and literacy; secondary education and technical training; higher education and research.

430 KOBAYASHI, Fumio
Development of the 'scientific experiment' movement in China and the problem of training scientific and technical manpower. *Ajia Keizai,* 6(9), 90-4, 1965.

431 LURCAT, Francois
Les Universites du Peuple. *La Recherche,* 3(23), 416-19, 1972.
The author describes technical training at Tsinghua University; medical education at Sun Yat-sen Institute.

432 MENDELSSOHN, Kurt
Aspects of science in China. *Arts and sciences in China,* 1(2), 26-9, 1963.
History of China's contacts with international science in Manchu period; as a result of opium war; under Communist regime. Training of Chinese scientists is relieving shortage of teaching and research manpower. 'In spite of their acute shortage of scientists, the Chinese have resisted the temptation of shortened university courses, specialization at school and other forms of educational pressure-cooking.' Vocational science training conducted in specialized training and research institutes.

433 MENDELSSOHN, Kurt
Science in China. *New Scientist,* 8(208), 1261-3, 1960.
Education in China; universities remain centers of scholarly rather than vocational studies; engineering and medicine taught at specialized institutes. 'Research fields nearer to practical application have first call.'

434 *NEW SCIENTIST*
Chinese colleges to re-open? 'Notes on the news', 39(611), 368, 1968.
Brief review of *New York Times* report of Chinese investigation report on Shanghai machine tools plant.

435 *NOTES ET ETUDES DOCUMENTAIRES*
L'enseignment dans la Republique populaire de Chine, No. 3197. Paris, Secretariat General du Gouvernement, Direction de la Documentation, 4 Jun 1965.
Chapter on higher education. Scientific research, P. 21-2.

436 *NOTES ET ETUDES DOCUMENTAIRES*
La formation des cadres scientifiques et techniques en Republic populaire de Chine (1949-1963), No. 3576. Paris, Secretariat General du Gouvernement, Direction de la Documentation, 28 Mar 1969.
The training of scientific and technical cadres in the PRC. Official politics and the domain of science and technology. The training of scientific personnel. Employment and utilization of scientific and technical cadres. Soviet aid: Soviet scientific and technical personnel in China; Chinese studying in the USSR; Exchange of scientific and technical information; Importance of Soviet aid. Western influence: Scientific cooperation between China and the West before 1949; Training of Chinese scientists and engineers in the West and in Japan; Consequences for Chinese science and technology; Future perspectives. Specialists in social sciences.

437 OLDHAM, C.H. Geoffrey
*Visits to Chinese universities*. Institute of Current World Affairs, unpublished newsletter CHGO-40, 8 Jan 1965.
General discussion of Communist party policy for higher education, entrance requirements, course work, job assignment, research, politics and the student, student life, staff condition, campus conditions, academic standards. Observations of laboratories in Nanking University, Fu Tan University, Hangchow Agricultural University, Sun Yat-sen University.

438 ORLEANS, Leo A.
Communist China's education--policies, problems, and prospects. *An economic profile of Mainland China*, U.S. Congress Joint Economic Committee. New York, Praeger, P. 499-518, 1968.

439 ORLEANS, Leo A.
Education and scientific manpower. 'Science in Communist China,' Gould. *AAAS*, No. 68, P. 103-27, 1961. Also *Education and Communism in China*, Stewart E. Fraser (ed.). London, Pall Mall Press, P. 363-87, 1971.
The education system. Conclusions about size and distribution of China's professional manpower based on educational statistics. Role of scientists and the organization of scientific research; reform of intellectuals.

440 ORLEANS, Leo A.
Medical education and manpower in Communist China. *Comparative Education Review*, Feb 1969. Also *Aspects of Chinese education*, Hu, C.T. (ed.). New York, Teachers College (Columbia University) Press, P. 20-42, 1969.
Higher medical education; list of higher medical institutions; higher medical manpower; secondary medical education and manpower;

lower medical personnel, spare-time medical training, part-time medical personnel, the Cultural Revolution.

441 ORLEANS, Leo A.
*Professional manpower and education in Communist China,* National Science Foundation, No.NSF-61-3, 1961. Also *Education and Communism in China,* Stewart E. Fraser (ed.). London, Pall Mall Press, P. 81-103, 1971.
II--Educational policies and problems. III--Primary and secondary education. IV--Higher education. V--Quality of education. VI--Science and technology. VII--Professional manpower. VIII--Survey of the population and labor force. Appendices include: C--Institutions of higher education. D--Institutions offering postgraduate courses. E--Specializations in higher technological institutions. I--Scientific research institutes in China.

442 PETROV, Victor P.
*China: emerging world power*. Princeton and London, Van Norstrand, 1967.
Technical training in China and abroad, P. 65-9.

443 PRICE, R.F.
*Education in Communist China*. London, Routledge & Kegan Paul, 1970.
Technical education, P. 143-4, 216.

444 RANGARAO, B.V.
Science in China (training and utilization). *Science and culture,* 32(7), 342-8, 1966.
Changing government and party attitude toward scientists; the education and training of scientists and engineers. List of references.

445 *SCIENCE NEWS LETTER*
China's science behind, 79(20), 315, 1961.
Brief article based on Leo A. Orleans, *Professional manpower and education in Communist China* (see entry No. 441).

446 *SCIENCE NEWS LETTER*
China's science growing, 89(6), 91, 1966.
Brief article on scientific manpower based on study by Chu-yuan Cheng (see entry No. 412).

447 SHARP, Ilsa
No ivory towers. *Far Eastern Economic Review,* 72(23), 64-6, 1971.
A new pattern in the educational system in China is emerging after the Cultural Revolution. Impressions based on the author's visit to China, chiefly Chungshan University. 'Chungshan University's curriculum shows a strong bias towards applied sciences . . . . The departments have been revamped to cover study in electronics, biology, synthetic materials, optics, dynamics, rare metals, politics, geography and Chinese . . . . The casualties of the Cultural Revolution were the departments of physics, maths, chemistry, foreign languages and philosophy.'

448 SHIH, Ch'eng-chih
The status of science and education in Communist China and a comparison with that in USSR. *Communist China Problem Research Series* EC-30. Hong Kong, Union Research Institute, 1962.
First presented as a paper at the Third International Sovietological Conference, 1960. 'It was intended to present my observations, through the angle of international relations, on Communist China's education in science and technology and on the difference between Communist China and the Soviet Union in bringing up their future generations to be 'new men of Communism' each according to her own historical and cultural background.' Contents: The traditional problems of education, science and technique; full utilization of existing educational, scientific and technical conditions; short courses in higher education; planning and progress in scientific and technical research; 'leap forward' in education and science; comparison of education and science in Communist China with their counterparts in USSR; the making of a 'new man of communism' in Communist China and USSR.

449 SWETZ, Frank
Training of mathematics trachers in the Peoples' Republic of China. *American Mathematical Monthly*, 77(10), 1097-1103, 1970.
'The present principal influence in Chinese education is the Communist Party and its doctrines as set forth in the writings of Chairman Mao Tse-tung . . . . In the sphere of education, 'walking on two legs' has resulted in the formation of numerous spare-time and part-time schools at various levels to supplement the expanded regular school system . . . . Despite the existence of these bogus 'schools', there are many types of bona fide education institutions in existence in the Peoples' Republic of China . . . . The burden of supplying trained teachers for these schools has fallen upon the traditional training institutions . . . . Three levels of teacher training institutions exist at present in Mainland China: the junior normal school, the senior normal schools, and the higher normal school or university.' Curricula in teacher training institutions.

450 TSANG, Chiu Sam
*Society, schools and progress in China.* Oxford, Pergamon, 1968.
'The scheme which the present author uses is not one of his choice, but was suggested to facilitate comparison among the countries under study.' 6--The school system; Research, P. 206-12. 7--Social, technological and international change.
9--Summary and conclusion; Scientific and technological achievements, P. 310-11.

451 U.S. CONGRESS: SENATE
Committee on government operation: subcommittee on national security staffing and operations. *Staffing procedures and problems in Communist China.* Washington, U.S. Government Printing Office, 1963.
Chap. V--Manpower and education; The labor force; The education system; Graduate work; Use of Soviet schools; Scientific manpower.

452 WANG, Chi
*Mainland China organizations of higher learning in science and technology and their publications.* Washington, Library of Congress, 1961.
Includes learned societies, universities and colleges, Academia Sinica and affiliated research institutes, Chinese Academy of Medical Science and branch institutes, Chinese Academy of Agricultural Sciences and branch institutes, governmental research organizations, and libraries. Publications include serial publications, abstracting and indexing services, bibliographies, and dictionaries. Information supplied for each entry includes official name of the organization in Romanized transcription, in Chinese characters, and in English; location or address; founding date; officials; research activities and facilities; and principal publications. Also, membership totals and local branches of learned societies, scientific departments of universities and colleges, size and type of collections of libraries and botanical gardens.

453 YAMADA, Keiji and TETSURO, Nakaoka
Science suggests China's direction. *The educational revolution uniting science and labor,* Vol. 6, Chap. IV. Tokyo, Mainichi Newspaper Co., Japan and China, 1972.
The 'unity of productive labor and education' surpasses ideology, does not degenerate into mere vocational education. It is a reversal of 'systematic education'. The basic form, 'combination of three', exists in all factories and permits the 'wisdom of the masses' to become science, reversing the relationship between teachers and students. Criticism of Dewey. The leading character of science. Ch'ien Hsueh-sen; theoretical physics.

454 YAMAGUCHI, Tomio
Medical education in the People's Republic of China. *Kagaku Asahi*, No. 3, P. 91-6, 1964. Trans. *JPRS* xx268, 24 Apr 1964.
Author's observations on the state of medical science and health in China while a member of the Japan-China Medical and Pharmacological Exchange Group visiting China in November 1962 at the invitation of the Chinese Medical Society.

455 YAO, York Bing
*Bibliography of the study on scientific training in institutions of higher learning in Communist China, 1958-1964.* New York, Columbia University, unpublished mimeo, n.d.
Bibliography divided into secondary sources: bibliographies; background readings; works for general reference; and primary sources; periodicals; newspapers and news agencies; Chinese Communist official documents and speeches; articles. List of articles in secondary sources. Chinese articles in translated sources.

See also the following entries:

Technology Policy:
Technology and Economic Growth. 1, 20, 26.
Transfer of Technology. 44.
Innovation and the Direction of Technological Change. 72, 92, 94, 95, 97, 98.
Policy for Science. 117, 119, 121, 123, 125, 129, 139, 146, 147, 152, 158, 163, 164, 165, 166, 167, 172, 176, 182, 184, 185, 188, 189, 190, 191, 192, 193, 194, 195, 196, 199, 206, 210.
Policy Toward Science. 218, 230, 234, 244, 248.
Research and Development:
Mathematics. 258, 260.
Physics. 261, 262, 268.
Astronomy. 269.
Biology. 277, 281.
Geology and Earth Sciences. 289, 292.
Geophysics. 301.
Agriculture. 304, 306.
Military/Space Exploration and Travel/Nuclear Energy. 321, 326, 327, 337, 359.
Popularization of Science. 458.
Agricultural Technology. 466, 467, 485.
Industrial Technology:
Electric Power Industry. 498.
Chemical Industry. 532.
Machinery Industry. 590.
Transport Equipment Industry. 596, 597.
Basic Metals Industries. 599.
Engineering. 631.
Medicine and Public Health. 685, 686, 687, 688, 690, 695, 702, 704, 706, 713, 714, 721, 725, 726, 729, 731.

## E Scientific and Technical Education at the Primary and Secondary Levels; Extension of Scientific and Technical Information; Popularization of Science

456 *CHINA NEWS ANALYSIS*
Secondary education, No. 147, 7 Sep 1956.
Between primary and university. The 1951 education reform. Number of students. Changing policies. Technical schools.

457 LEAR, John
How Red China is taking science to its peasants. *Saturday Review*, P. 45-6, 6 Mar 1965.
Based on interview with C.H.G. Oldham, concerning peasant-scientist Chen Yung-kang. 'The Chinese Communist rulers know that to be a modern power the superstitions and intuitive outlook of most Chinese people must be replaced by science and rational thought.'

458 OLDHAM, C.H. Geoffrey
Science and superstition. *Far Eastern Economic Review*, 48(1), 14-18, 1965.
Description of science education in primary and middle schools . . . visited by the author in 1964. 'The aspects which I found most interesting was the efforts which are being made to replace superstition and intuitive judgements at the peasant level with science and rational thought.'

459 OLDHAM, C.H. Geoffrey
*Visits to Chinese schools*, Institute of Current World Affairs, unpublished newsletter CHGO-39, 6 Jan 1965.
Description of Soochow Experimental Primary School; Shang Ming Middle School (Shanghai), Peking No. 15 Municipal Middle School, Soochow Senior Middle School. Teaching of science in middle schools.

460 RICHMAN, Barry
Economic development in China and India: some conditioning factors. *Pacific Affairs*, 45(1), 75-91, 1972.
Qualitative comparison of the environments of Chinese and Indian economic performance. Conclusion: China seems to lead in overall economic development and in creation of an environment conducive to substantial and sustained economic growth. Conditioning factors: educational aims, methods and accomplishments (China seems to have done a better job of matching its overall educational system with the manpower requirements of economic development and of effectively using skilled manpower); the socio-cultural sphere; attitudes toward scientific methods (partial and erratic success in popularizing the scientific method, sacrificed to ideological extension); the view toward risk-taking--the political and legal environment; defense and military policy; foreign policy (China's high degree of self-sufficiency achieved at the cost of waste, inefficiency and self-sacrifice); political stability and organization; banking, fiscal and monetary policy, taxation; organization of capital

markets; factor endowment; population; market size and export capability; social overhead capital.

See also the following entries:

Technology Policy:
  Technology and Economic Growth. 24.
  Innovation and the Direction of Technological Change. 69, 78, 97, 98.
Policy for Science. 115, 122, 132, 137, 167, 171, 172, 176, 183, 210.
Policy Toward Science. 254.
Research and Development:
  Agricultural. 304.
Scientific Communication. 367.
Higher Education and Technical Training. 403, 429, 430, 439, 441.
Agricultural Technology. 471.
Population Control. 735, 737, 739, 740, 741.

# F Testing and Standardization

461 *CHINA NEWS ANALYSIS*
Symbols in science, No. 366, 7 Apr 1961

462 *CHINA NEWS ANALYSIS*
Weights and measures: industrial standards, No. 427, 6 Jul 1962.

See also the following entries:

Policy for Science. 210.

# SECTION V

## *Technology in China*

The entries in this final section refer to technology as it is used in various sectors of the Chinese economy and society. In the absence of precise data from China, policy objectives and the degree of success in implementing development policies must be deduced from information of the kind listed here: mostly general descriptions of specific industries and enterprises; of agricultural production; social policies such as employment, health care, population control and environmental control; defence and military strategies; and programmes of international technical assistance—in all of which, technology and technological advance are a means to the attainment of policy objectives. Each of these general descriptions refers *inter alia* to technology in a specific context, but only a few focus specifically on the questions of technological change, the choice of techniques, the direction of technological change, or the sources of technological innovation, which are the focal questions in the study of science, technology and development. Thus, in a sense, this section describes the outcome of Chinese policies for science and technology.

The demand for new technologies is the sum of demands arising in various sectors of the economy, plus those for various social, political, and military purposes. Even though these demands may be complementary and interdependent, their total may exceed the capacity of available R & D resources and/or the developing country's access to foreign technologies. Thus one purpose of science and technology policies, closely reflecting the objectives of overall development goals, is to assign some order of priority to these different technological demands.

For example, the industrialization process depends not only on the development of new technologies of industrial production, but also on new sources of raw materials, new sources and forms of energy, and on the development of efficient transportation and communication facilities. In addition, agricultural production as a form of capital accumulation, must be increased in order to supply raw materials to industry, feed industrial workers, and provide export goods which will earn foreign exchange. Agricultural development, too, requires new technologies of production, and hence imposes demands on scientific resources for R & D, the extension and diffusion of new agricultural techniques, and for technological innovation in the industries which manufacture capital equipment for agriculture. Furthermore, the first objective for most developing countries is to substitute locally manufactured goods for imported goods in the domestic market, and then, to open up international markets for the products of the developing country's own industries. Therefore, international trading patterns as well as domestic needs impose their own set of demands for technological advance.

New technologies are also required for certain social purposes which do not immediately or directly contribute to economic growth, but which are part of the development process. Some social welfare problems have been inherited from pre-industrial conditions, while others have occurred as a consequence of initial industrialization. For example, improvement of medical technologies and the extension of public health services may raise the general health and physical well-being of the population to unprecedented levels. Even where traditional medical and health practices are highly developed, modern scientific medical techniques may still improve upon traditional techniques. Yet in many developing countries, modern health and medical techniques and more productive technologies have enabled larger populations to live at subsistence level, rather than improved the welfare of a fairly stable population. Depending on how 'over-population' is interpreted by the planners, they may direct part of the scientific research effort toward the development of appropriate techniques of population control and their dissemination and diffusion throughout the population.

Industrialization has had other deleterious effects on the social fabric of developing countries. Notably, the growth of industry has had far-reaching consequences for patterns of employment, and for the institutional structures which absorbed labour in the pre-industrial society. The growth of industry in many developing countries has attracted labour from the agricultural sector, and displaced labour employed in traditional handicraft industries in greater numbers than can be absorbed in the comparatively small industrial sector. The problem is exacerbated by the importation of capital-intensive industrial technologies, which further minimize the number of industrial work places by virtue of their high investment costs. Furthermore, in the developing country industry tends to be geographically concentrated near sources of raw materials or near transportation facilities, or is located in certain areas for historical reasons, such as the establishment of foreign-owned enterprises in treaty ports in China. This may lead to mass migration to the cities and the consequent overburdening of urban facilities.

The social consequences of the use of modern industrial technologies must be weighed in the planners' choice of technology. Their conclusion may be that labour-intensive technologies should be used for social objectives in order to generate employment, and that small-scale plants should be constructed in order to facilitate dispersion of industrial development, absorb labour in rural areas and so avoid the uncontrolled growth of urban population. The problem in science policy, then, is to direct research toward the development of labour-intensive, small-scale technologies which still use the factors of production efficiently and produce a sufficiently large surplus output to finance further technological advance.

'Development' implies increasing control over nature. Thus 'environmental control' refers to control over destructive natural phenomena, such as droughts and floods. The use of modern technologies in the developing countries may however result in some of the same destruction of the natural environment as has occurred in the industrialized countries. Thus, the choice of technology and the development of new technologies for use in the developing countries should ideally take into consideration the environmental impact, as well as the social and economic implications of a particular technology, and new techniques should be developed to counteract the detrimental effects of current technologies on the environment.

China is probably unique among developing countries in being a donor of technical assistance, rather than a recipient. Such aid, of course, does not directly benefit the donor country itself, but is intended to serve more immediate, usually political, purposes. The type of technology offered, and the terms on which it is made available to other countries, may be indicative of the donor country's own development and of its own experience of technical assistance, as well as of its international political strategies. Note the distinction between *technical* and *economic* (financial) assistance, which is not included in this category.

A demand for new technologies arises not only from specific economic and social needs, but also for political and military purposes. A nation's research efforts may be devoted to the development of new military technologies or to advanced scientific achievements which are deemed to enhance its national prestige. The proportion of R & D effort so directed depends on the structure of the science policy-making system and its responsiveness to demands arising in military and political quarters. Despite the potential spin-off of economically- and socially-useful technologies, and despite the expansion of scientific facilities which may occur under these circumstances, such projects may be regarded from the development perspective as subtracting from the total research effort which might otherwise be expended on the development of appropriate, efficient, economic and social technologies.

# A Agricultural Technology

463 BROADBENT, K.P. (ed.)
The development of Chinese agriculture 1949-1970. *Commonwealth Bureau of Agricultural Economics: Annotated Bibliography No. 3*, n.d. Supplement A, 1960-1970, Dec 1971.
Bibliography of Chinese and non-Chinese books and articles on various aspects of agriculture in China.

464 BUCK, J. Lossing
Fact and theory about China's land. *Foreign affairs*, 28(1), 92-101, 1949. Also *Bulletin of the atomic scientists*, 6(12), 365-8, 1950.
China's fundamental problem is pressure of population upon resources. General farm mechanization is impractical without changes in the man-land ratio, or until machinery is available at low prices and until there is a cheap source of power. Through improved technology, the size of farm business could be increased; with reform of farm tenancy, credit and taxation, China could increase food production by 50%. Contributions of American agriculturalists in China. 'Collective farming and redivision of land would probably decrease production per capita . . .'

465 CHAO, Kang
*Agricultural production in Communist China*, 1949-1965. Madison, Milwaukee and London, The University of Wisconsin Press, 1970.
Pt. 2--Inputs utilization and technological changes. Pt. 3--Some general aspects of technological transformation. Pt. 4--Agricultural mechanization and improvement of farm implements.
Pt. 5--Irrigation and rural electrification. Pt. 6--Fertilization. Pt. 7--Cropping systems and breeding.

466 CHENG, Tien-hsi
Insect control in Mainland China. *Science*, 140(3564), 269-77, 1963.
Published works on entomology in China deal, by and large, with applied research; limited basic research of commendable quality has been accomplished by senior Chinese entomologists trained in America or Europe. Chemical control of insect pests; biological control using parasites or predators; traditional cultural control practices further developed through modern technology; mechanical and physical controls using irrigation and light traps; insect forecasting and reporting. Research on migratory locusts.

467 CHENG, Tien-hsi
Production of kelp--a major aspect of China's exploitation of the sea. *Economic Botany*, 23(2), 215-36, 1969.
'. . . considerable advances have been made in Mainland China in aquicultural practices . . .' Growth of technical manpower, research and educational institutions specializing in algology and related disciplines. Economic and social impacts resulting from promotion of kelp farming. Utilization of kelp as food and as source of industrial raw materials. Lines of research in kelp production.

468 CHINA RESEARCH INSTITUTE
Reform of agricultural technology in China. *China materials monthly report*, No. 73, Mar 1954.

469 DAWSON, Owen L.
China's two-pronged agricultural dilemma: more chemical fertilizer and irrigation needed for more food. *Current Scene*, 3(20), 1-13, 1965.
Brief review of background of Chinese agriculture; analysis of China's potential ability to use chemical fertilizer and water.

470 DAWSON, Owen L.
*Communist China's agriculture, its development and future potential.* New York, Praeger, 1970.
This study uses available provincial data which . . . help present a fuller picture of the national economy; analyzes productive factors as a basis for estimates of potential supply for home consumption and export; and analyzes political prospects. 6--Agricultural research and development, P. 92-111. 7--Chemical fertilizer requirements, P. 112-34. 8--Power resources for agriculture, P. 135-69.

471 ERISMAN, Alva Lewis
China: agricultural development, 1949-1971. *People's Republic of China: an economic assessment*, U.S. Congress, Joint Economic Committee. Washington, U.S. Govt. Printing Office, P. 112-46, 1972.
Two distinct periods of agricultural development policy are distinguished by investment policies. 1949-61: investment from within agriculture itself, intensive application of labor on an inelastic supply of cultivated land. 1962-71: large and increasing amounts of chemical fertilizers, pesticides, equipment for irrigation and drainage projects, farm machinery. First period: slow pace in technological change. Inputs under the new strategy: water conservation and high yield fields; research; extension service; new varieties of seed; rural electrification; agricultural tools and machinery; tractors; pesticides and fungicides; chemical fertilizers. Tabulated statistics.

472 GOODSTADT, Leo
China mounts a war on pests. *Far Eastern Economic Review*, 75(7), 27, 1972.
Evidence of insect damage to agricultural harvests in 1971 suggests that modernization of agricultural techniques, using hybrid wheat and rice strains and multiple-cropping, and shortage of modern pesticides has upset the natural balance in China. Attempt to control insect pests and plant diseases using modern technology--bacteriolysis, chemical insecticides and drugs produced from micro-organisms, aerial afforestation, artificial catalytic rain-making. National Conference on Agricultural Mechanization in 1971 planned for total farm mechanization by 1975, but has resulted in policy disputes.

473 HINTON, William
*Iron Oxen: a documentary of revolution in Chinese farming.* New York, Monthly Review Press, 1970.
'Postscript 1970': '*Iron Oxen* was written as a recollected journal of the author's life and work in China during 1949 and 1950. It describes the first steps in the creation of a system of mechanized state farms as a critical turning point in the Chinese revolution . . . .' The revisionist view: socialist forms of agriculture cannot be developed until industry is able to provide tractors, fertilizers, and insecticides and the schools can provide a large cadre of technicians, scientists, etc. Mao Tse-tung's view: a fundamental shift from individual to collective agricultural production relations can provide a market for growing industries, supply industry with raw materials and can make possible the introduction of modern techniques into the Chinese countryside.

474 ISHIKAWA, Shigeru
Long-term outlook for Chinese agriculture. *Long-term prospects for the Chinese economy,* Pt. IV(2), Chap. VI, Shigeru Ishikawa (ed.). Tokyo, Asian Economics Research Institute, No. 172, 1967.

475 ISHIKAWA, Shigeru
Recent changes in Chinese agricultural techniques. *Ajia Keizai,* 3(1), 14-21, 1962.

476 JONES, P.H.M.
Creeping modernization. *Far Eastern Economic Review,* 46(7), 350-2, 1964.
Application of science to agriculture. Official Chinese policy distinguishes between 'ordinary farmland' and 'land destined to produce "good harvests irrespective of drought and excessive rainfall and provide stable, high yields".'

477 JONES, P.H.M.
Machines on the farm. *Far Eastern Economic Review,* 45(11), 479-81, 1964.
Agricultural mechanization.

478 JONES, P.H.M.
One million tractors. *Far Eastern Economic Review,* 45(10), 431-3, 1964.
Agricultural mechanization.

479 KAWAMURA, Yoshio
Supply of and demand for lumber and afforestation policy in China. *Ajia Keizai,* 7(9), 69-82, 1966.
I--Summary of forestry resources. II--Demand for and supply of timber. III--Forestry policy.

480 KAWAMURA, Yoshio
Technological revolution in Chinese agriculture. *Ajia Keizai,* 6(9), 65-77, 1965.

481 KUMASHIRO, Yukio
On reformation of agricultural technology during the Great Leap Forward period. *Ajia Keizai,* 9(12), 2-15, 1968.

482 KUO, Leslie T.C.
Agricultural mechanization in Communist China. *The China Quarterly*, No. 17, P. 134-51, 1964. Also *Industrial development in Communist China*, Choh-ming Li, P. 134-50. New York, Praeger, 1964.
The Chinese Communists believe that by using the combined methods of old and new, native and foreign, scientific and primitive, mechanical and manual, the motive power for agricultural production can be increased at a higher speed.

483 KUO, Leslie T.C.
Industrial aid to agriculture in Communist China. *International Development Review*, 9(2), 6-10, 29, 1967.
'The most important objective of industrial aid to agriculture is to speed up the technical transformation of agriculture.' Four areas of concentration in the "aid agriculture" program are mechanization, electrification, water conservation and chemicalization.

484 KUO, Leslie T.C.
*The technical transformation of agriculture in Communist China*. New York, Praeger, 1972.
'This study is primarily concerned with the technical aspects of agricultural development in Communist China. In the following chapters, the overall policies, guiding principles, programs, and methods for the technical transformation of agriculture adopted during the first two decades of the Communist regime will be examined . . . followed by a discussion of the major technical measures for improving agricultural production . . . . Primary attention will be given to the planning and execution of the programs rather than to the details of agricultural technology. In each case, the situation before the Communist takeover, the highlights of the programs carried out by the new regime, and the major accomplishments and problems will be observed. On the basis of such a perspective the progress of the technical transformation agriculture will be evaluated, and an attempt will be made to determine, *inter alia*, the extent to which the technical innovation has helped agricultural productivity in Communist China and the prospects for the next few decades.'

485 LARSEN, Marion R.
China's agriculture under Communism. *An economic profile of Mainland China*, U.S. Congress Joint Economic Committee. New York, Praeger, P. 197-267, 1968.
Education and technology in agriculture, P. 227-30; irrigation and water conservancy, P. 239-43; fertilization, P. 243-6; mechanization, P. 247-50.

486 LETHBRIDGE, Henry
China: collectivization and mechnization. *Far Eastern Economic Review*, 39(7), 311-12, 1963.
Agricultural mechanization.

487 LETHBRIDGE, Henry
Tractors in China. *Far Eastern Economic Review*, 39(12), 616-19, 1963.

488 LETHBRIDGE, Henry
Trends in Chinese agriculture. *Far Eastern Economic Review*, 40(9), 499-500, 1963.
Application of science to agriculture.

489 LIU, Jung-chao
Fertilizer application in Communist China. *The China Quarterly*, No. 24, P. 28-52, 1965.
Role of chemical fertilizer in Chinese agricultural policy; production; imports; consumption; investment, cost of production, and prices of chemical fertilizers; yield response to fertilizer; gains from fertilizer application.

490 PERKINS, Dwight
*Agricultural development in China 1368-1968*. Chicago, Aldine Publishing Co., 1969.
'During the millenia prior to the founding of the Ming dynasty in A.D. 1368, China's population spread into the rice-growing regions surrounding the Yangtze River and from there further southward into present-day Kwangtung. This great migration required a whole new technology fundamentally different from that used on the dry millet and wheat lands of the North China plain.' Of the rise in grain output between 1398 and mid-twentieth century, about half is attributable to a four-fold increase in the amount of land cultivated; the remaining half was the result of a rise in grain yields per acre, within a stagnant technology. 'Traditional' agricultural technology is not sufficient to support China's population. In the early 1960s, 'traditional' technology was given a secondary role and policy focused on large-scale investment in modern capital inputs: electrification, modern farm implements, machinery, and chemical fertilizer.

491 PORCH, Harriet E.
The use of aviation in agriculture and forestry in Communist China. *RAND Paper* P-3566, Apr 1967.
Agricultural aviation is well established in China . . . but a shortage of resources means that activities may be spread so thin as not to be very effective. Use of aircraft in agriculture and forestry. Organization of civil aviation in Communist China. Types of aircraft in use.

492 RICHARDSON, S. Dennis
A modern-day Marco Polo visits China. *American Forests*, P. 6-15, 47-52, Feb 1965.
Forestry in China; observations based on a visit to China in 1963.

493 TAMURA, Saburo
Chemicals and mechanization processed by the mass movement of workers and peasants. 'Science and technology in modern China'. *The Asahi Asia Review*, 3(1), 1972.

494 YAMAMOTO, Hideo
Agricultural revolution and farm production methods in China. *Ajia Keizai*, 5(3), 20-32, 1964.
II--Basic characteristics of methods of agricultural production.
III--Indigenous innovation, collectivization and agricultural

production techniques. IV--Collective enterprise and changes in overall production capacity. V--The change of traditional production methods.

495 YAMAMOTO, Hideo
*The process of change in the agricultural technology system in China.* Tokyo, Asian Economics Research Institute, No. 122, 30 Oct 1965.
I--Basic nature of agricultural technology system. II--Analysis of the strength of agricultural producers: irrigation; fertilizers; improvement of product varieties; mechanization.
III--Changes in the character of agricultural labor.

See also the following entries:

Technology Policy:
  Technology and Economic Growth. 1, 3, 4, 9, 24, 26.
  Choice of Techniques. 31, 32, 37.
  Innovation and the Direction of Technological Change. 67, 85, 106, 107, 113.
Policy for Science. 138, 171.
Research and Development:
  Biology. 278.
Scientific Communication. 374, 393, 394.
Industrial Technology:
  Electric Power Industry. 497.
  Electronics Industry. 508.
  Chemical Industry. 522, 524, 525, 526, 527, 528, 529.
  Machinery Industry. 576, 580, 581, 582, 587.
Natural Resources. 656.
Environmental Control. 751.

# B Industrial Technology

## 1 Electric Power Industry

496 ASHTON, John
Development of electric energy resources in Communist China. *An economic profile of Mainland China,* U.S. Congress, Joint Economic Committee. New York, Praeger, P. 297-316, 1968.
I--Electric energy resources in Communist China: hydroelectric resources; coal; petroleum. II--Organization of the electric power industry. III-- Development of the electric power industry: capacity, investment, and generation; transmission networks; general technological level; production of electric power equipment. IV--Consumption of electric energy. V--Rates, revenues, costs, and profits.

497 CARIN, Robert
Power industry in Communist China, *Communist China Problem Research Series* DC44. Hong Kong, Union Research Institute, 1969.
I--Background. II--Condition of motive power resources: coal; water power; petroleum; natural gas; atomic energy. III-Capital construction: volume of investment; scale of aid from the Soviet Union and East European countries; geographical distribution; construction time; construction cost. IV--Power generation. V--Power transmission. VII--Rural electrification.
VIII- Conclusions.

498 CHIEN, Yuan-heng
A study of the electric power industry on the Chinese mainland. *Chinese Communist Affairs,* 4(4), 21-31, 1967.
Power resources are abundant; Chinese have expended great efforts to train technicians, improve technology and manufacture power generating and transmitting equipment. Loss of Soviet aid replaced with West European equipment.

499 MUZAKI, Shotaro
Analysis of the present state of China's electric power industry. *China Materials Monthly Report,* No. 131 (Tokyo) Feb 1959.
I--Outline of the Chinese electric power industry. II--Chinese economic development and electric power requirements. III-China's power construction and its level of development seen in the hydro-electric and steam-generated department.

500 ONOE, Etsuzo
Energy policy: the pattern of development in the 1970s. 'Science and technology in modern China'. *The Asahi Asia Review,* 3(1), 1972.

501 ONOYE, Etzuso
The Chinese electric power industry. *Long-term prospects for the Chinese economy,* Pt. I, Chap. VII, Shigeru Ishikawa (ed.). Tokyo, Asian Economics Research Institute, No. 76, 1964.

502 ONOYE, Etzuso
The Chinese power industry (Supplement by N. Akabane). *Long-term prospects for the Chinese economy*, Pt. II, Chap. IX, Shigeru Ishikawa (ed.). Tokyo, Asian Economics Research Institute, No. 102, 1966.

503 YEH, K.C.
Electric power development in Mainland China; prewar and postwar. *RAND memorandum* RM-1821, 27 Nov 1956.
Major trends in the electric power industry, 1932-1955, focusing on 1932-36 and 1949-55. Sect. II--Output and allocation of electric power in Mainland China: close relationship between industrial production and electric power output, 1950-55. III--Electric power generating capacity in Mainland China. IV--Investment in the electric power industry in the Five Year Period 1953-57 . . . . Expansion of the industry is hampered by inadequate supply of steam turbines and of technical manpower. V--Institutional and technical development.

See also the following entries:

Technology Policy:
Choice of Techniques. 40.
Innovation and the Direction of Technological Change. 67, 92, 97, 103.
Policy for Science. 156, 171.
Research and Development:
Military/Space Exploration and Travel/Nuclear Energy. 320, 345, 351, 357.
Agricultural Technology. 470, 471, 483, 490.
Industrial Technology:
Petroleum Industry. 536, 537, 542.
Machinery Industry. 576, 578, 579.
Mining Industry. 622.
Engineering. 634, 637.
Transportation and Communications. 644.
Natural Resources. 650, 652, 656.

## 2 Electronics Industry

504 AKAGI, Akio
Large-scale electronic computers for general use. 'Science and technology in modern China'. *The Asahi Asia Review*, 3(1), 1972.

505 AUDETTE, Donald G.
Computer technology in Communist China, 1956-1965. *Communications of the ACM*, 9(9), 655-61, 1966.
Initial planning, organization and education aspects of computer technology and automation; machine development progress, with Soviet aid and all Chinese made; computer applications; trend of automation--production process control rather than data processing; 'yun ch'ou hsueh', 1958-60. China has a marginal computer capability.

506 *ELECTRONIC DESIGN*
Red Chinese are turning out computers with 'modest' ICs, Vol. 20, 28 Sep 1972.
Interview with American computer experts following 'recent' visit to China. Technical details of computers seen: about a dozen. '. . . we saw most of the computers in China.' 'The use of computers in China is generally limited to scientific and engineering calculations . . . virtually no process control with computers.'

507 *ELECTRONICS INTERNATIONAL*
China poised for 'Greap Leap' into the forefront of science, 24(1), 251-2, 1969.
Article based on report by Jon Sigurdson for Swedish Royal Academy of Sciences (entry No. 191). The Chinese computer industry; development of process-control equipment for chemical and petrochemical plants; telecommunications. Chinese monitor scientific developments abroad, import advanced instruments.

508 GOODSTADT, Leo
From the land, a new power struggle. *Far Eastern Economic Review*, 75(4), 42-3, 1972.
Debate in China over the role of the electronics industry in economic development. Technical change must be accompanied by social revolution, although Mao insists on exploiting the advances of modern science. 'The battle over advanced technology (represented by electronics) and technical innovations achieved by 'self-reliance' is far from over.' Agricultural mechanization.

509 KIRK, Don
China's three way stretch. *Electronics*, P. 129-35, 1967.
Three-fold purpose of Chinese policies for developing the electronics industry: to make electronics serve strategic nuclear projects; to produce prestige wares and consumer goods; and to produce broadcast communications equipment for propaganda use.

510 LÜBECK, Lennart
Electronics in today's China: some impressions from the visit of the Minister of Industry. *Teknik och Industri i Kina*. Stockholm, Ingenjörsvetenskapakademien Report 44, 1972.

511 *NEW SCIENTIST*
China's progress in computers. 'Notes on the news', 41(632), 111, 1969.
Brief review of article in *Electronics International* by John (sic) Sigurdson, 6 Jan 1969 (entry No. 507).

512 NYBERG, P. Russell
Computer technology in Communist China. *Datamation*, Feb 1968.
Organization of R & D in computer technology; Soviet aid; Chinese computers. China still lags in development of computer technology but is narrowing this gap; computer technology will continue to be subordinated to supporting military science in China; it is doubtful whether the Chinese 'large transistorized digital computer' represents any kind of a technical breakthrough.

513 REICHERS, Philip D.
The electronics industry of China. *People's Republic of China: an economic assessment,* U.S. Congress, Joint Economic Committee. Washington, U.S. Govt. Printing Office, P. 86-111, 1972.
Description of history, production facilities and level of technology of the electronics industry, by product group: electronic components; electronic instruments; computers; communications equipment; consumer entertainment equipment; military electronics. Appendix: major facilities in China's electronics industry.

514 SATO, Masumi
Course of development of the electronics industry. *Long-term prospects for the Chinese economy,* Pt. IV(2), Chap. VIII, Shigeru Ishikawa (ed.). Tokyo, Asian Economics Research Institute, No. 172, 1970.

515 SATO, Masumi
Technological development in China viewed through the electronics industry: an engineer's view. *The developing economies,* 9(3), 315-31, 1971. Also *Ajia Keizai,* 12(12), 2-14 (in Japanese), 1971. 'Our overall thesis is that Chinese technology has been built up over the years from a foundation of basic science to the point where China is now approaching the levels of Western Europe and Japan in theoretical fields.' Although design technology in the manufacturing industries has reached a high level, there are problems in materials and operations technology. Thus the potential for technological development centers on design technology.

516 TSU, Raphael
High technology in China. *Scientific American,* 227(6), 13-17, 1972.
Impressions from the author's (a solid-state physicist) visit to China in summer 1972. '. . . the Chinese have recognized that modern industry is grounded in advanced technology . . . . There is high technology in China, including specifically the computers, control systems and instrumentation that make a modern industrial society function.' Brief description of attainments in conventional industry: visit to Peking oil and chemical complex; machinery on view at the Shanghai Industrial Exhibit. High-technology attainments: Chinese copies of and variations of Russian jet aircraft; nuclear weapons; rocketry; semi-conductor technology; the electronics industry: R & D. The Institute of Semi-conductors of the CAS; foreign publications in electronics received in China; topics of research. University/industry contacts at Peking university. Computers. 'One of the aspects of Chinese industry that impressed me is the evidence of local autonomy under general guidelines established by the government.'

517 *WASHINGTON SCIENCE TRENDS*
Mainland China computers, 28(22), 129, 1972.
Report of observations by T.E. Cheatham, 'a leading U.S. computer specialist', on his visit to China. '. . . mainland China is producing second and third-generation computers on an assembly-line basis, using their own integrated circuits.' Chinese

scientists are aware of international research; publications; self-reliance 'means that the U.S. may not have much of a computer market on the mainland.'

518 YAMAGUCHI, John
Electronics in Communist China. *Electronics,* Vol. 33, P. 32-3, 23 Dec 1960.
The Chinese electronics industry may reach the goal set in the second FYP of 'catching up with world standards' by 1962-63. The groundwork was laid during the first FYP on Russian, Hungarian and East German technological foundations. Review of research program and accomplishments. Chinese military technology copies Soviet models, concentrates on industrial and consumer electronics components and communications.

See also the following entries:

Technology Policy:
  Innovation and the Direction of Technological Change. 67, 97, 103.
Policy for Science. 138, 156, 157, 159, 160, 173, 200.
Research and Development:
  Electronics. 271, 272.
  Military/Space Exploration and Travel/Nuclear Energy. 318, 340.
Industrial Technology:
  Machinery Industry. 576, 589.
  Basic Metals Industries. 602.
Transportation and Communications. 639.

# 3 Chemical Industry

519 AKABANE, Nobuhisa
Chemical industry: technological self-reliance preceded the development stage, 'Science and technology in Modern China'. *The Asahi Asia Review*, 3(1), 1972.

520 AKABANE, Nobuhisa and KOJIMA, Reiitsu
China's chemical industry. *Long-term prospects for the Chinese economy*, Ishikawa, Shigeru (ed.), Pt. II, Chap. VII. Tokyo, Asian Economics Research Institute (The Institute of Developing Economies), No. 102, 1966.

521 *CHINA NEWS ANALYSIS*
Plastics; synthetic fibre. No. 541, 20 Nov 1964.

522 CLOSE, Alexandra
Down to earth. *Far Eastern Economic Review*, 50(10), 517-22, 1966. Production of chemical fertilizers. Choice of production scale.

523 DALYELL, Tam
Chemical industry in China today. *Chemistry & Industry*, No. 1, P. 10-11, 1 Jan 1972.
The author's observations of the Shanghai chemical plant. Production processes: production of sulphuric acid for fertilizer production; coal is the basis for production of ammonia and methanol; new drying rotary kiln for the urea process; conversion of sulphur dioxide into sulphuric acid using 'crudish' form of water spraying. Organization of the plant: 5 main workshops, 3 auxiliary workshops, maintenance workshop. Levels of output employment. Innovations on display at the Shanghai Industrial Exhibition. China as an exporter of technology. Efforts to control pollution, urban drift.

524 HSIAO, Chi-jung
Production and supply of chemical fertilizers on the Chinese Mainland. *Issues and studies*, 1(6), 41-57, 1965.
Annual output. Construction and production capacity of its fertilizer industry. Annual supply. Conclusion.

525 KAMBARA, Shu (ed.)
*The chemical industry in China*. Tokyo, The Institute of Developing Economies, Mar 1970.
(In Japanese) 1--General discussion: objectives and problems in research on the Chinese chemical industry; conditions of development of the chemical industry--international comparisons and China's special characteristics; industrial production and the chemical industry; commodity trade of the chemical industry. 2--Specific discussions of the chemical industry: the soda industry; the fertilizer situation (including the technical movement in the Wuching fertilizer factory); textiles; chemical fibre industry; plastics; rubber; dyestuffs, paint, cosmetics; agricultural chemical industry. 3--Related production: chemical

plant engineering; petroleum industry. 4--Condition of development of Chinese science and technology. 5--Special characteristics of the Chinese chemical industry and future development: minutes of a discussion.

526 MACDOUGALL, Colina
Fertilizer drive. *Far Eastern Review*, 49(1), 14-16, 1965.
Old factories are being extended for fertilizer production as well as being supplemented by new plants. Small-scale industry meets immediate requirements. Chinese claim to make ten different fertilizers, plus urea.

527 MACDOUGALL, Colina
Fertility rites. *Far Eastern Economic Review*, 42(2), 5, 1964.
Italy has three contracts for fertilizer plants; existing plants; more output needed although short-term targets met.

528 NETRUSOV, A.A.
Development of the chemical industry in the Chinese Peoples' Republic. *Byulleten' tekhniko-ekonomicheskoi informatsii*, No. 1, P. 75-8, 1959. Trans. in *L.L.U. Translation Bulletin*, P. 1-5, Jul 1959.
Description of achievements in developing the chemical industry during the first Five-Year Plan, and plans for 1958 and following. 'The first Five-Year Plan (for 1953-7) envisaged founding a powerful chemical industry as one of its most important tasks (with special emphasis on mineral fertilizers, acids, soda, dyes, rubber and pharmaceuticals). During the years of the plan 31 plants were built or reconstructed, and of these 14 were running. The Soviet Union assisted in building 10 of these plants.'

529 WILSON, Dick
Chemicals for the communes. *Far Eastern Economic Review*, 44(11), 533-5, 1964.
Visit to Wuching Chemical Works. History of chemical fertilizer production.

530 WILSON, Dick
Plastics in Shanghai. *Far Eastern Economic Review*, 45(1), 16-17, 1964.
Visit to Shanghai chemical factory.

531 YOUNG, G.B.W.
Mainland China's chemical industry. *RAND Memorandum* RM-4504-PR. Santa Monica, The RAND Corporation, Mav 1965.
Monograph reviewing growth of Chinese chemical industry, 1949-1964; import of complete chemical plant and technical know-how; prospects for indigenous development of chemical technology.

532 YOUNG, G.B.W.
Red China. *Chemical engineering progress*, 61(12), 37-40, 1965.
The chemical industry in 1949; growth of the chemical industry since 1949: production increase, 1953-1957; industry expansion, 1958-1964; related foreign trade. Current technical status of the chemical industry: a 1959 review of the chemical industry by

the vice-minister of the chemical industry suggested a 'strong and growing chemical producing industry, a weak chemical converting industry, and a promising chemical processing industry.' 'The announced 1959 capability was being demonstrated in 1964.' Three factors suggest a very promising outlook for the technical status of the chemical industry in the near future. First, there is the increasing number of graduates . . . and of matured technical personnel . . . . Second, . . . great emphasis on and presumably large outlays for science and technology. And third, there is the import of technical know-how which is to accompany the increasing import of plants, particularly chemical, from Western Europe and Japan.' Tabulated statistics: production outputs in 1952 and 1957, and 1957 target; distribution of China's imports with time; complete-plant orders from Western Europe and Japan.

See also the following entries:

Technology Policy:
Technology and Economic Growth. 3, 26.
Choice of Techniques. 40.
Innovation and the Direction of Technological Change. 97, 103, 110, 111.
Policy for Science. 138, 156.
Agricultural Technology. 465, 469, 470, 471, 472, 482, 489, 490, 493, 495.
Industrial Technology:
Electronics Industry. 507, 516.
Petroleum Industry. 541.
Textile Industry. 573, 574.
Machinery Industry. 576, 578.
Mining Industry. 615, 624, 627.
Engineering. 635.
Trade. 658, 668.
Medicine and Public Health. 714.

## 4 Petroleum Industry

533 CHEN, Ke-Chung
Petroleum industry in the Chinese Mainland. *Chinese Communist Affairs*, 1(3), 9-18, 1964.
Basic principles for development of petroleum industry: development of natural and synthetic petroleum production; large, medium and small-scale production; combination of modern foreign technology and old native methods. Organization; Soviet aid; major oil fields; major refineries; synthetic oil plants and oil shale deposits; petroleum resources and explorations; production and consumption; impact of oil shortage on the economy.

534 CHIMIE ET INDUSTRIE--GENIE CHIMIQUE
*L'industrie petroliere en Chine populaire*, 101(1), 86-9, 1969.
Petroleum resources in China are very limited. Total production is estimated as 10 million tons in 1966, of which an important

part is schist petroleum. Research and production; refining; transport and consumption.

535 ESSO STANDARD S.A.F.
Departement information. *L'industrie petroliere en Chine populaire. Informations Economiques: Bulletin Interieur* No. 437, P. 6-8, 20 Jan 1968.
Research and production; refining; transport and consumption. Map.

536 GARDNER, Frank J.
Chinese oil flow up, but much larger gains needed. *The Oil and Gas Journal,* 69(50), 35-9, 1971.
Chinese oil production has recovered strongly since the Cultural Revolution. Refining capacity being steadily expanded. Exploration activity locating oil reserves. Estimate of production, based on *Peking Review* article, November 1971. '. . . a significant step-up in imports of both petroleum and oil industry equipment will occur during the remainder of this decade.' 'With nuclear-energy development still in its infancy, coal will probably remain China's principal fuel source for industry power generation, rail transport, and heating during the foreseeable future.'

537 HAO, Paul L.C.
The Chinese Communist petroleum industry. *Chinese Communist Affairs,* 4(1), 20-26, 1967.
Development of petroleum industry; import of equipment from Soviet Union, from Italy, Japan and France. Outlook for petrochemical industry. Petroleum as energy supply.

538 HEENAN, Brian
China's petroleum industry. *Far Eastern Economic Review,* 49(13), 565-7, Pt. I, 1965; 50(2), 93-5, Pt. II, 1965.
I--China's petroleum industry is now capable of considerable technological achievement, from exploration through production phases to refining. Soviet assistance. Plant imports. Research. Exploration for reserves. II--Development of the petroleum industry. Innovations.

539 HEENAN, L.D.B.
The Chinese petroleum industry. *Tijdschrift voor Economische en Sociale Geografie,* Vol. 57, P. 149-60, Jul/Aug 1966.
China's petroleum industry surveyed on the basis of secondary and primary reports . . . 'is now capable of considerable technological achievement affecting most branches of the industry, ranging from exploration through the various production phases to refining. Reserve estimates have been increased substantially in the post-1949 period and significant expansion has been reported and probably achieved in the output of both crude and synthetic oils. The refining sector has also developed markedly in both processing capacity and range of products manufactured.' Technological resource base: Russian assistance included training of Chinese; provision of Russian experts; supervision of exploration, well-drilling, extraction and refining; provision of plant and equipment; introduction of new methods and skills. After initial

setback following cessation of Russian aid, China's petroleum industry has made much progress. Manufacture of refining equipment: China is developing self-sufficiency, but due to technical deficiencies, substantial imports are still necessary. Scientific research in oil institutes and refining centers. The physical resource base; synthetic oil reserves; petroleum production; petroleum imports; the expanding refining industry; distribution of refineries; self-sufficiency and oil exports, future growth and some development problems: imbalance between distribution of reserves and production centers and location of effective demand.

540 HEENAN, L.D.B.
The petroleum industry of monsoon Asia. *Pacific Viewpoint*, 6(1), 65-95, 1965.
Section on China, P. 89-95: the industry's resource base: 'much exploration . . . has brought positive results . . . the Soviet Union and other Communist countries have supplied equipment and skilled personnel.' Crude oil production, 1943-1962: 'levels of crude oil production and output of all oils make Communist China second only to Indonesia as monsoon Asia's most important oil-producing nation.' Imports and oil consumption: production of all oils in 1962 was sufficient to fulfil only some two-thirds of domestic demand . . . between 1950 and 1960 most of China's imported oil came from Soviet Russia; since 1960, imports from Rumania compensate for falling levels of import from the Soviet Union. Refining: large refineries (Lanchow; Fushun) and small refineries, an attempt to overcome difficulties of transportation. Tabulated statistics: oil reserves, production, imports and consumption in China (1959-1962).

541 *THE JAPAN ECONOMIC JOURNAL*
Peking building huge petrochemical center, 21 Sep 1971.
Report of a Japanese correspondent on visit of delegation from Japanese Liberal-Democratic Party to new complex under construction southwest of Peking. Refinery is a part of petrochemical complex with installations, plants and residences occupying 500.000 sq. meters; will refine 2,500,000 tons of crude oil annually, produce benzole and lubricating oil, and manufacture synthetic rubber, textiles, plastics and other chemical products. 'High degree of Chinese technology in the petrochemical field' noted by the visitors.

542 LETULLIER, Andre
Le petrole en Chine populaire. *Revue Francaise de l'Energie*, No. 188, P. 199-209, Mar 1967.
General geology of China; petroleum basins; bituminous schists; resources of petroleum; Chinese production; reserves; imports and consumption; refineries; transport; petroleum as a source of energy.

543 MACDOUGALL, Colina
China keeps the oil flowing. *The Times review of industry and technology*, 5(8), 58-9, 1967.
Effect of the Cultural Revolution on petroleum production.

Taching, Karamai, and Yumen fields. Chinese economy still relies mainly on coal, but oil is essential for economic modernization. Imports of crude from Soviet Union ceased in 1961, imports of petroleum products fell to 38,000 tons in 1965; upward trend in Chinese exports of petroleum products. Output of refined increasing faster than output of crude, indicating bottleneck in refining facilities may have been solved. Refining facilities at Lanchow, Shanghai, and Taching.

544 MEYERHOFF, A.A.
Developments in Mainland China, 1949-1968. *The American Association of Petroleum Geologists Bulletin,* 54(8), 1567-80, 1970.
Based on publications from Russia, Japan, Taiwan and Mainland China, plus some unpublished data. Stages of petroleum exploration, 1949-1968. Generalized geology of Mainland China's hydrocarbon basins: stratigraphy, structural settings of basins. Exploration progress, 211 B.C.-1968 A.D. Production, pipelines and refining. Future of Mainland China's petroleum industry. Tables: 4--Data on principal Mainland China oil fields, Jan 1 1969. 5--Mainland China gas fields, Jan 1 1969. (Both listing basin, field name and number, latitude and longitude, year discovered, location basis, producing depth, producing fm, ages, density and/or gravity, production in 1968, cumulative production; ultimate recovery). 6--Principal Mainland China refineries and topping plants.

545 MIKI, Ken'ichiro
Communist China's petroleum situation. *Ajia Keizai,* 5(2), 2-11, 1964.
Based on various Western publications.

546 *THE OIL AND GAS JOURNAL*
China blames Russia for oil troubles, 61(52), 49, 1964.
Withdrawal of Soviet experts, denial of sets of equipment, cuts in Soviet exports to China upset development of oil industry.

547 *THE OIL AND GAS JOURNAL*
China nurses flow back up to '66 level, 68(26), 28-9, 1970.

548 *THE OIL AND GAS JOURNAL*
Chinese oil hurt by 'Revolution', 66(50), 144, 149, 1968.

549 *THE OIL AND GAS JOURNAL*
Chinese Reds claim catalytic reformer built from scratch, 65(4), 62, 1967.

550 *THE OIL AND GAS JOURNAL*
To develop an oilfield, get rid of ghosts and ogres, 64(48), 126, 130-1, 1966.
Technology used at Taching oilfield.

551 *THE OIL AND GAS JOURNAL*
Red China battles for oil status, 65(8), 33-6, 1967.
Based on study by L.D.B. Heenan (entry No. 539). Reserve estimates

increased by exploration, 1949-late 1950s, when large scale development started on several new fields. Refining sector has developed in processing capacity and range of products. Development of Chinese oil technology from 1949 to 1960 achieved largely through substantial help from Russia; Chinese trained in USSR; secondment of Soviet personnel to China. Chinese now claim to be able to make relatively sophisticated equipment and instruments, engineering capabilities, but advanced design and other technical deficiencies persist. Reserves; new discoveries; exploration; current production. Synthetic oils; natural gas; coal. Oil imports. Refining has developed into a major industry. The Chinese have established a technological base.

552 *THE OIL AND GAS JOURNAL*
Red China near self-sufficient in oil, 68(40), 78-9, 1970.
Based on A.A. Meyerhoff in the *Bulletin of the American Association of Petroleum Geologists*, Aug 1970 (entry No. 544).

553 *THE OIL AND GAS JOURNAL*
Red China reports higher production, 67(5), 126-7, 1969.

554 *THE OIL AND GAS JOURNAL*
Red China reports Taching output up, Vol. 68, P. 27, 1970.

555 *THE OIL AND GAS JOURNAL*
Taching oil hikes Red Chinese output, 67(16), 62-3, 1969.
Brief article on technical facilities at the Taching oilfield. Production appears to be from shallow low-pressure pay . . . . Rigs appear small and of primitive design. Low substructure indicates absence of high-pressure blowout prevention equipment. First Chinese designed drilling rig (1968) powered by diesel units totalling 3000 hp; weight about 300 metric tons. Taching construction began April 1962; atmospheric and vacuum distillation unit and thermal cracking unit operating by November 1963. All equipment is said to be Chinese-made, 'but it's considered likely that units of other Chinese refineries were cannibalized to provide some of the parts for the new facility.'

556 *THE OIL AND GAS JOURNAL*
Turmoil cripples Red Chinese oil, 66(22), 73-4, 1968.
Based on study by the information department of Esso Standard S.A.F., 1968 (entry No. 535). Soviet deliveries of drilling equipment 'jumped' in 1968. Esso study finds 24 oil and 11 natural gas fields in production. Design capacities of principal crude oil refineries.

557 *PETROLEUM PRESS SERVICE*
China's response, P. 88-90, Mar 1965.
'Over the years, China's indigenous production has become continuously more significant as a source of oil supply for the local market, and the point has now been reached where it is officially claimed that the country is "basically self-sufficient".' The Chinese also claim to have achieved basic self-sufficiency in refining. Indigenous and imported refining technology. Imports of crude and of petroleum products in the 1960s. Forecasts

of the Chinese market for petroleum products; prospects of Chinese exports.

558 *PETROLEUM PRESS SERVICE*
Openings in China, 39(10), 359, 361, 1972.
Reduction of Chinese imports of oil. Refining capacity is growing but cannot keep pace with crude output; hence China's offer to exchange crude for refined products with Japan. Japanese, American, Canadian and Iranian drilling technology may be imported by the Chinese.

559 SLEZAK, F.
China: Entwicklung der Erdölindustrie (Development of the petroleum industry). *Osterreichischen Geographischen Gesellschaft.* Mitteilungen, 112(2/3), 422-5, 1970.
Development of the petroleum industry. Petroleum reserves and geological characteristics. Table of output in 1907, 1939, 1943, 1949, 1956, and 1968. Soviet technology replaced after 1960 by special equipment from Rumania and Western countries.

560 TRETIAK, Daniel
Has China enough oil? *Far Eastern Economic Review,* 44(11), 536-7, 1964.
Chinese oil industry being developed to meet essential domestic requirements without large imports from abroad.

561 *L'USINE NOUVELLE*
La Chine, producteur de petrole . . . d'apres-demain? (China, petroleum producer . . . the day after tomorrow?), No. 18, P. 36-7, 2 May 1968.
Research and production; survey and exploration; five refineries in action (Lanchow; Touchangtzou; Dairen; Tsinghai; Nanking); transport and consumption.

562 WESTGATE, Robert
Red China claims large oil resources are being developed. *World Oil,* 151(6), 138, 144, 146-8, 153-4, 157-8, 1960.
Increase in oil production over ten years; exploration. Oil fields and refineries. Production of equipment for the oil industry and of geophysical instruments in China. Supply of equipment and technical know-how from Soviet Union, Rumania, Czechoslovakia, Eastern Germany and Hungary. Scientific research for the oil industry in Institute of Petroleum, CAS; Petroleum Research Institute; Academy of Geological Research of the Ministry of Geology is concerned with theory and technical problems; Institute of Geology of the CAS stresses theoretical research.

563 WILSON, Dick
Fuel from Fushun. *Far Eastern Economic Review,* 44(8), 374-5, 1964.
Visit to Fushun Oil Refinery and Lung Fung coal mine.

564 *WORLD OIL*
International outlook issue: China, 153(3), 215-16, 1961;

155(3), 250, 1962; 157(3), 202, 1963; 159(3), 182, 1964; 161(3), 200, 202, 1965; 163(3), 192, 1966; 165(3), 172-4, 1967; 167(3), 221, 1968; 169(3), 213, 1969; 171(3), 184, 1970; 175(3), 132, 1972.
Annual reviews of the petroleum industry in China.

565 *WORLD OIL*
Red China's 'lost' oil field is found, 166(1), 9, 1968.
Taching oil field . . . is reported by Nationalist Chinese sources to be 80 miles northwest of Harbin, Manchuria, near the Anta railroad station. Production is from cretaceous sandstones, 2600-4000 feet, on a large anticline. About 800 wells have been drilled.

566 *WORLD PETROLEUM*
China's crude output rises to meet needs, 41(11), 16, 1970.
Based on A.A. Meyerhoff (entry No. 544).

567 WORLD PETROLEUM EDITOR
Mainland China. *World Petroleum*, 42(11), 23, 1971.
'How much crude oil and/or other hydrocarbons are being produced in Mainland China? How much of it is likely to become an import advantage for that country? Or, conversely, is there a shortage of energy, and is China potentially a place for others to send their export oil?' Economics, politics and diplomacy will dominate decisions on further exploration or purchase of crude oil in the world market.

568 WORLD PETROLEUM REPORT
China. *World Petroleum*, Vol. 7, P. 276, 1961; Vol. 8, P. 34, 1962; Vol. 9, P. 111, 1963; Vol. 10, P. 110, 1964; Vol. 11, 1965; Vol. 12, P. 110, 1966; Vol. 13, P. 87-8, 1967; Vol. 14, 1968;
Brief annual reviews of developments in petroleum industry in various countries, including China. (1964 and 1967: no oil activities reported).

569 YEH, K.C.
Communist China's petroleum situation. *RAND Memorandum* RM-3160-PR. Santa Monica, The RAND Corporation, May 1962.
Production, imports, and consumption of petroleum in Communist China, 1949-60. Prospects of expanding indigenous production and reducing dependence on imports from the Soviet Union.

See also the following entries:

Technology Policy:
- Choice of Techniques. 40.
- Innovation and the Direction of Technological Change. 67, 92, 97, 103.

Policy for Science. 138.
Industrial Technology:
- Electric Power Industry. 497.
- Electronics Industry. 507, 516.
- Chemical Industry. 525.

Industrial Technology:
Machinery Industry. 576.
Mining Industry. 615, 622, 623, 624, 625, 627, 628, 629, 630.
Engineering. 635.
Natural Resources. 650, 651, 652, 654, 656.
Trade. 668.

# 5 Textile Industry

570 BRENNAN, Charles
China: 'King' cotton. *Far Eastern Economic Review*, 54(12), 648-59, 1966.
China is not near the level of the world's most advanced textile industries, as measured by the machines used.

571 *FAR EASTERN ECONOMIC REVIEW*
Technical license for China, 38(4), 259-60, 1962.
Kurashiki Rayon Co. license for Chinese manufacture of vinylon.

572 KOJIMA, Reiitsu
Textile industry. *Long-term prospects for the Chinese economy*, Pt. IV(2), Chap. IX, Shigeru, Ishikawa (ed.). Tokyo, Asian Economics Research Institute, No. 172, 1967.

573 TSUTSUMI, Shigeru
From natural to man-made fibres in 'Science and technology in modern China'. *The Asahi Asia Review*, 3(1), 1972.

574 WILSON, Dick
Black crow into peacock. *Far Eastern Economic Review*, 44(13), 633-4, 1964.
Visit to Shanghai textile mill. Import of textile machinery. Synthetic fibres.

See also the following entries:

Technology Policy:
Choice of Techniques. 40.
Innovation and the Direction of Technological Change. 97, 103.
Industrial Technology:
Chemical Industry. 521, 525, 528.
Transportation and Communications. 644.
Trade. 658, 668.

# 6 Machinery Industry

575 CHAO, Kang
*The construction industry in Communist China*. Edinburgh, Edinburgh University Press, 1968.
Pt. I--Scope and organization. Pt. II--Output. Pt. III--Inputs and

technological aspects. Chap. II--Technological aspects; Quantity and quality of designing work; Mechanization; Standardization and prefabrication. Conclusion: 'the main weakness of the Chinese construction industry has been its low technical level. The majority of construction workers are young and new in their profession. Designers and architects have not been equipped with the modern technology necessary for handling the designing of industrial complexes and other large civil engineering projects. In the early years they relied heavily on the technical consultation of Russian experts and on designing materials imported from the Soviet Union . . . . As far as technical development in the construction industry is concerned, emphasis has been placed throughout the period on standardization of building materials and structural parts, prefabrication, use of standard designs, and reduction in the number of models; these procedures have effected a significant savings in cost. More important, these measures provide an effective remedy for the shortage of designing staff and skilled construction workers in China.'

576 CHENG, Chu-yuan
*The machine building industry in Communist China*. Edinburgh, Edinburgh University Press, 1972. Also Growth and structural changes in the Chinese machine-building industry. *The China Quarterly*, No. 41, P. 26-57, Jan/Mar 1970. Also The effects of the Cultural Revolution on China's machine-building industry. *Current Scene*, 8(1), Jan 1970.
'The fundamental object of this study is to present a comprehensive picture of the machine-building industry in China: demand and supply, imports and exports, growth rates in gross and net output value, input-output relationships, changes in output composition, and the industry's contribution to national goals. The approach is mainly empirical and fact-finding.' 'The study also examines some technical aspects of the industry, the composition of its output, and its sectoral structure. The significance of these aspects . . . since the industry provides capital goods for other industries and thus is a major factor in productivity, the technical level of its products bears vital consequences for overall industrialization.' Comparison of China's achievements in the machine-building industry from 1952 to 1966 with those in advanced countries in terms of . . . technical levels. 'The Chinese industry is technologically far behind the advanced countries' . . . in number of types of machinery produced; capacity of machinery produced; and in degree of precision.

577 CHINA RESEARCH INSTITUTE (Tokyo)
Present conditions of the machine industry in China. *China materials monthly report*, No. 89, Jul 1955 (in Japanese).

578 CHINA RESEARCH INSTITUTE (Tokyo)
Technical development of Chinese heavy industry. *China materials monthly*, No. 114, Aug 1957.
1--Development of heavy industry after the founding of the Communist Republic: General state of development of the iron and steel industry; general state of development of the machine tools industry; general state of development of the chemical

industry; development trends in the nature and technology of heavy industry. 2--Heavy industry's problems seen in technology: The automotive industry; ball bearing industry; construction equipment industry; electric plant industry. Appendix: China's capacity to build various kinds of machines.

579 ECONOMICS RESEARCH INSTITUTE
Analysis of present state of industrial technology in China (2) (in Japanese). *Machine industry economic research report*, 42-10. Jun 1968.
Classification of types of machine industry; introduction of Soviet technology; general machine classification and present state; electrical machinery; precision machinery.

580 HARASHINA, Kyoichi
Growth centered around tractor building in 'Science and technology in modern China'. *The Asahi Asia Review*, 3(1), 1972.

581 KOJIMA, Reiitsu
The agricultural machinery and implement industry. *Long-term prospects for the Chinese economy*, Pt. III, Chap. VII, Shigeru Ishikawa (ed.). Tokyo, Asian Economics Research Institute, No. 119, 1967.

582 KOJIMA, Reiitsu
Agricultural machinery and tools industry in the development of a self-sustaining national economy. *Ajia Keizai*, 7(9), 45-68, 1966 (in Japanese).
I--The distinction of production of agricultural tools from village side-line enterprise. II--The relation between agricultural machinery and tools and foodstuffs. III--The road to Chinese-style agricultural mechanization.

583 KOJIMA, Reiitsu
The Chinese machine tool industry. *Long-term prospects for the Chinese econcmy*, Pt. II, Chap. VIII, Shigeru Ishikawa (ed.). Tokyo, Asian Economics Research Institute, No. 102, 1966.

584 MUNTHE-KAAS, Harald
China's mechanical heart. *Far Eastern Economic Review*, 48(9), 398-400, 429, 432, 1965.
Development of the Chinese machine building industry.

585 MUZAKI, Shotaro
Development of the Chinese machine industry in the past ten years. *China materials monthly report*, No. 143, Feb 1960 (in Japanese).
I--Overview. II--Machine material industry's development: by department of the machine industry. III--Increase in labor productivity in the machine tools industry.

586 MUZAKI, Shotaro
The technical level of the Chinese machine tools industry. *China research monthly*. Tokyo, China Research Institute, No. 201, Nov 1964 (in Japanese).

I--What the problem is. II--Present conditions of the Chinese machine tools industry. III--Major potential and characteristics of Chinese manufacture of important machine tools. IV--Examination of the level of technical development of the Chinese machine tools industry in comparison with Japan.

587 ONOYE, Etsuzo
Production of agricultural machinery in China. *Ajia Keizai*, 3(10), 40-60, 1962 (in Japanese).
II--Structure of production of agricultural implements: Before 1957--national enterprises, local factories, handicraft production; 1958-60--the direction of factories, production of agricultural machinery in non-specialized factories, construction of new factories, finance, technology, materials.

588 *PRODUCT ENGINEERING*
Red China industry nears technical independence from USSR, Vol. 32, P. 15-16, Feb 1961.
Survey of China's machine-building industry, based on the paper by E.K. Nieh for the American Association for the Advancement of Science. 'It is perhaps of utmost significance that China has apparently mastered these various industrial techniques to the point where she is no longer dependent on Soviet aid and can even promise assistance to lesser developed countries in Asia.'

589 *PRODUCT ENGINEERING*
Red China joins USSR to pioneer new automation techniques, Vol. 29, P. 23-4, Nov 1958.
Computer control technology in the USSR and, inter-alia, in China. 'Chinese universities are cooperating with machine-tool plants and government to forge a concept of numerical control that is unparalleled in the West . . . . Chinese sources . . . envision a single master tape with a control program directing scores of lathes that operate simultaneously. Such a concept could conceivably swamp the world with products whose production costs include almost nothing for skilled manpower.'

590 STUDY GROUP FOR SCIENCE AND TECHNOLOGY OF NEIGHBOURING NATIONS
Foreign machine industry survey committee. *Analysis of present status of Communist China's science and technology*. Tokyo, Jul 1965 (in Japanese. Trans. *JPRS* 41, 859).
A compilation of facts obtained by about ten experts of industry and business during a tour of China in 1964(?) presented in outline form. Chap. 1--The technological framework of Communist China; the technology related to industrial materials; machine tools and precision machines; types of equipment produced; technical management; indices of productivity; plant organization; concrete manifestations of technical policies; training of technicians; level of technicians; comparison with Japan; shortcomings in their technology; absence of parts industry; imported machinery alone is insufficient; deficiencies in technical knowledge; outdated technology; regional development of combinats; research organization; imported facilities. Chap. 2--Status of technology in Communist China.

591 WILSON, Dick
No more by line of skin. *Far Eastern Economic Review*, 44(12), 595-7, 1964.
Shenyang number one machine tool factory.

592 WILSON, Dick
Presses from Shenyang. *Far Eastern Economic Review*, 44(7), 337-9, 1964.
Shenyang heavy machine plant achieves independence of foreign technical assistance and importation of foreign equipment.

593 WILSON, Dick
Shanghai machine tools. *Far Eastern Economic Review*, 44(5), 241-3, 1964.
Observations made by the author on his visit to the Shanghai machine tools factory.

594 YAMATO, Shuzo
Present and future of the machine tool industry in 'Science and technology in modern China'. *The Asahi Asia Review*, 3(1), 1972.

See also the following entries:

Technology Policy:
Choice of Techniques. 32, 40.
Innovation and the Direction of Technological Change. 67, 80, 85, 92, 97, 103, 110, 111.
Policy for Science. 138, 157, 171.
Scientific Communication. 393.
Agricultural Technology. 465, 471, 477, 478, 483, 485, 486, 487, 490.
Industrial Technology:
Electric Power Industry. 497, 498.
Electronics Industry. 516.
Petroleum Industry. 539, 549, 551, 555, 562.
Mining Industry. 624.
Engineering. 634.
Trade. 658, 659, 667.

# 7 Transport Equipment Industry: Motor Vehicles

595 *AUTOMOTIVE INDUSTRIES*
China's liberation model truck goes into production, 115(6), 66-9, 1956.
Brief comment on production of first trucks of all-Chinese manufacture. Photos of equipment in use at No. 1 motor car plant, Changchun.

596 WESTGATE, R.
Industrialization of China moves ahead. *Automotive Industries*, 118(9), 45-8, 1958.
China, after starting production of motor vehicles more than a

year ago, is now making jet aircraft in quantity and has begun trial manufacture of tractors. Machine tool production. Chemical, radio, meter and electrical engineering industries have been developed rapidly during five years on concentration on heavy industry. Soviet aid a key factor in China's rapid industrialization; 156 large-scale industrial projects are the core of this aid. Chinese engineers and technicians learn plant design, construction and operation at the Soviet-assisted projects. Training of Chinese engineers, administrators and skilled workers in the Soviet Union. Import of Soviet equipment; Chinese-designed plant and equipment.

597 WESTGATE, R.W.
Industrialization behind the bamboo curtain. *Automotive Industries*, 113(11), 66-8, 106, 122, 1955.
Review of the automotive industry in China. Changchun No. 1 and No. 2 car plants are part of the 156 key projects in Soviet aid program: Soviet experts and technicians have been associated with every stage of construction of the plant, and Soviet engineers will supervise starting up of the plant. Contribution of Chinese industry to equipping the Changchun plant: steel, machine tools. Training of skilled manpower in the Soviet Union and at three technical schools in Manchuria. Manufacture of tractors and aircraft.

See also the following entries:

Technology Policy:
  Innovation and the Direction of Technological Change. 103.
Industrial Technology:
  Machinery Industry. 576, 578.
Transportation and Communications. 641, 646.
Trade. 658.

# 8 Basic Metals Industries

598 AKENO, Yoshio
The Chinese iron and steel industry. *Long-term prospects for the Chinese economy*, Pt. I, Chap. VI, Shigeru Ishikawa (ed.). Tokyo, Asian Economics Research Institute, No. 76, 1964.

599 ALLARD, Blain and Riviere
*Mission Siderurgique Francaise en Chine 16 Oct-7 Nov 1965*. Paris, Institut de Recherches de la Siderurgie.
Metallurgical Research Institute of Peking; Metallurgical factory at Chekingchan; Institute for Metallurgical Training, Peking; Metallurgical plant at Anshan; Metallurgical Society's Research Center at Anshan; No. 1 Metallurgical plant of Shanghai.

600 CLOSE, Alexandra
China streamlines her steel. *Far Eastern Economic Review*, 47(9), 365-8, 1965.

Survey of the Chinese iron and steel industry. Rumors of Chinese imports of plant and machinery may be substantiated in 1965. Technological development in 1964.

601 ERSELCUK, Muzzaffer
The iron and steel industry in China. *Economic Geography*, No. 4, P. 347-71, Oct 1956.
Conclusion: The Chinese steel industry is developing rapidly . . . (which) could scarcely have taken place except under the peculiar economic system prevailing in China, a system under which the government is in a position to foster the expansion of industries despite natural or economic disadvantages. Russian aid in the form of equipment, technical knowledge, and capital have also aided in the promotion of the Chinese iron and steel industry. Review of China's major resources of raw materials: iron ore, coal, scrap, other raw materials. Description of production facilities and technology: Anshan, Penchihu, Shihchingshan, Tientsin, Luan (Tangshan), Paotou, Taiyuan, Paochin, Shanghai/Kiangsu, Maanshan, Hankow, Chungking, Kunming.

602 *FAR EASTERN ECONOMIC REVIEW*
A new priority for steel, 74(40), 26, 31-2, 1971.
Guidelines in the fourth FYP. Re-emphasis on iron and steel vs. electronics technology as the basis of China's industrial revolution.

603 HSIA, R.
Anshan steel. *The China Mainland Review*, 1(3), 12-23, 1965.
Review of the development of the Anshan Iron and Steel Corporation.

604 HSIA, R.
China's key steel bases. *The China Mainland Review*, 2(2), 103-12, 1966.
'The present article attempts to evaluate the importance of the Anshan, Wuhan and Paotow steel centers in China's steel industry in terms of capacity and output.' Attention is focussed on iron-smelting, steel-refining and rolling; concentration on a few bench-mark years.

605 HSIA, R.
Paotow steel. *The China Mainland Review*, 2(1), 10-15, 1966.
History of the establishment of the Paotow Iron and Steel Corporation. Description of technical features.

606 HSIA, R.
Wuhan steel. *The China Mainland Review*, 1(4), 16-23, 1966.
History of the Wuhan Iron and Steel Corporation; description of technical features.

607 HSIA, R.
The development of Mainland China's steel industry since 1958. *The China Quarterly*, No. 7, P. 112-20, Jul/Sep 1961.
Local, small scale steel production units; indigenous technology in 1958 mass campaign. In the short run, the steel campaign was

a waster of economic resources. In the long run, the campaign has left a definite imprint on the further development of China's steel industry.

608 KOJIMA, Reiitsu
The iron and steel industry. *Long-term prospects for the Chinese economy*, Pt. IV(2), Chap. VII, Shigeru Ishikawa (ed.). Tokyo, Asian Economics Research Institute, No. 172, 1970.

609 ONOYE, Etzuso
The Chinese iron and steel industry. *Ajia Keizai*, 3(2), Feb 1962.
Survey of the Chinese iron and steel industry from 1949. Anshan and Paotow Companies. Indigenous production in small iron and steel enterprises.

610 SEWELL, R.
A guide to the Chinese steel industry. *The British Steelmaker*, P. 259-63, Jul 1960.
Description of iron and steel technologies in China, based on the Russian technical press and some Chinese data. Iron ore reserves. Equipment and techniques; product and process innovations in the Chinese iron and steel industry. Research and development: science policy and scientific relations with the Soviet Union; the Peking Ferrous Metals Institute. Plant manufacture: most plant and equipment is built to Soviet design, but the Chinese attempt to adapt these designs to local conditions.

611 WILSON, Dick
Apricot-time in Anshan. *Far Eastern Economic Review*, 44(9), 471-2, 1964.
Visit to Anshan Iron and Steel Works.

612 WU, Yuan-li
*The steel industry in Communist China*. New York, Praeger, for The Hoover Institution, 1965.
Economic analysis of the iron and steel industry in Chinese development. Chap. 3--Growth of the iron and steel industry since 1949 including an analysis of the changes in production and capacity. Chap. 7--Locational development and its impact in 1949-60 including the 'backyard furnace' drive and location. Appendix A--Estimates of Mainland China's iron ore reserves. Appendix D--Changes in production technique, by Ronald Hsia including organization of metallurgical research.

See also the following entries:

Technology Policy:
  Choice of Techniques. 28, 40.
  Innovation and the Direction of Technological Change. 67, 97, 103, 110, 111.
Policy for Science. 173.
Research and Development:
  Military/Space Exploration and Travel/Nuclear Energy. 318.
Scientific Communication. 393.

Industrial Technology:
Electronics Industry. 516.
Machinery Industry. 576, 578.
Mining Industry. 615, 616, 620, 622, 623, 624, 625, 627, 628, 629.
Transportation and Communications. 644.
Trade. 658.

# 9 *Mining Industry*

613 DWYER, D.J.
The coal industry in Mainland China since 1949. *Geographical Journal*, 129(3), 329-38, 1963.
History of the development of the coal industry in the People's Republic. Conclusion: 'The years since 1949 have seen a remarkable development of the Chinese coal industry . . . it is clear that few of the current Chinese claims about the industry can be accepted without reservation.'

614 *FAR EAST TRADE AND DEVELOPMENT*
Modernizing coal production, P. 159-60, Mar 1969.
Compilation of reports on capital construction in Chinese coal industry.

615 GEALY, Edgar J. and WEI Anton W.T.
Mainland China. *Mining Annual Review* (published by *Mining Journal)*, P. 248-9, 251, 1965; P. 261-3, 1966; P. 270-3, 1967.
1965: Review of atomic development; Sino-Japanese trade; the oil industry; coal; iron and steel; base and minor metals in 1964. 1966: 'In 1965 the mineral industry of Communist China reflected much of the benefits of the previous retrenchment period which followed the debacle of the Great Leap. While production increases were generally modest, apparently considerable gains were made in terms of variety of products, quality standardization, rationalization of facilities, and completion of short range construction programs. More sober planning and the application of fundamental technology appears to have had the effect of firming the mineral and metal raw material base for agriculture and manufacturing industries.' Foreign trade; iron and steel; coal; petroleum; non-ferrous and other metals; fertilizers; salt. 1967: Considerable progress in China's mineral industry during 1966. Estimated output of steel, coal, crude oil, refined petroleum, chemical fertilizers. Trade: export of traditional Chinese metals, import of fertilizers, 'modest' tonnages of steel, crude and refined petroleum imports held to low levels 'because of China's successful efforts to exploit the recently discovered oil reserves and place in operation sufficient domestic refining capacity to supply the relatively modest needs of the country.' Iron and steel; petroleum; coal. (*Mining Annual Review* 1968, entry no. 628; 1969, entry no. 630).

616 HALPAP, P.
The mining industry in the Chinese People's Republic. *Steel & Coal*, 187(4979), 1218-20, 1963.
Translation of an article originally appearing in *Bergbautechnik*, 13 Feb 1963. Expansion of trained manpower has enabled systematic geological survey, which has revealed deposits of iron ore, bauxite, wolfram, molybdenum, lead, and manganese ores, which are now gradually being developed. Coal measures are well suited for modern winning methods; 55 large collieries, as well as small locally-worked collieries south of the Yangtse River; mechanization of production. Oil: intensive prospecting has disclosed deposits over 2.7 million sq. km; development of four petroleum fields (Chiushuan, Dzungaria, Tsaidam, Szechuan). Large iron-ore fields have been found in Kansu and Chinghai, which will enable north west China to develop as a center-point of the iron and steel industry. Other minerals: phosphates, gold, mica, asbestos, marble.

617 HEMY, G.W.
*The Chinese coal industry*. Warrington, Joseph Crosfield and Sons Ltd., 1961.
Pts. I-VI--Comprehensive survey of the Chinese coal industry: background, economics, organization, geographical distribution; mining methods, machinery, transport, open cast and hydraulic mining, coal preparation plants; safety and working conditions; coal industry research. Based entirely on Chinese and Soviet sources.

618 MARTYNOV, V.V.
The coal industry of China. *Byulleten' tekhniko-ekonomicheskoi informatsii*, No. 2, P. 84-5, 1959.
'Improvements in methods of coal-extracting have also had a great influence on the development of the Chinese coal industry . . . . Intensive work is also being carried out on the mechanization of the basic processes of coal working . . . . Chinese miners have mastered over a short period of time complex modern techniques . . . . The raising of the standard of mechanization of the basic production processes in coal output has been the basic factor that conditioned the growth of the productivity of labor.'

619 PEARSON, G.E.
Minerals in China. *Far Eastern Economic Review*, 36(10), 513-15, 1962.
Survey of coal, copper, lead, zinc, tin, aluminium, antimony, tungsten, manganese production.

620 SATO, Masumi
Mining industry, including the history and future of the iron and steel industry in 'Science and technology in modern China'. *The Asahi Asia Review*, 3(1), 1972 (in Japanese).

621 STEPHENS, Michael Dawson
The coal industry of N.W. and S.W. China. *Asian Review*, 1(2), 106-112, 1964.
History of the development of the coal industry in modern China.

'As yet coal mining in the interior of China is not as economical as within the eastern fields, but with the Communist policy of improving the transport network throughout the state and the establishment of a skilled labor force . . . the differences in production costs should later prove more equal. China is now one of the three great coal producers of the world and it seems likely that this industry will continue to hold a prime position, but with increasing emphasis on the little exploited resources of the interior.'

622 VAN RADERS, H.A.
De ontwakende reus (The awakening giant). *Boortoren en Schachtwiel*, 10(5), 94-9, 1965.
Mining, steel industry and energy in China. Based on material published in 'A concise geography of China' (Jen Yu Ti, Foreign Language Press, Peking Institute, 1964): material from the 'New Zealand-China Society'; Japanese and Hong Kong economic reports; radio broadcasts from Peking, Djakarta, Rangoon, and Tokyo in English and French; publications from China, Indonesia and Taiwan. Mineral resources; the future; iron and steel--survey of refineries, coal deposits; hydro-electric stations; oil; other minerals in China; maps.

623 WANG, K.P.
China. *Mining Annual Review* (published by *Mining Journal*), P. 405-9, 1970; P. 396-7, 399-401, 1971.
1970: 'A clear upturn in mineral and industrial activity in China was noted for 1969 . . .' Metal and mineral production--'a very good year . . . with the sharpest gain in petroleum.' Petroleum, coal, iron and steel, major base metals, export metals, nuclear activities, foreign trade. 1971: Review of the mining industry in 1970. 'History has shown that the Cultural Revolution not only held down production but impeded education and training as well, and the Sino-Soviet confrontation adversely affected the normal process of China's economic development. Conversely, the recent conciliatory attitude of the Chinese in domestic and international affairs may portend more realistic industrial programs in the future. Petroleum; coal; iron and steel; major base metals; export metals; nuclear activities; non-metals and fertilizers; foreign trade. (*Mining Annual Review*, 1965, 1966, 1967, entry no. 615; 1968, entry no. 628; 1969, entry no. 630).

624 WANG, K.P.
Mineral industry of Mainland China. *Bureau of Mines Minerals Yearbook*. Washington, U.S. Dept. of the Interior, P. 1107-20, 1965; P. 193-202, 1967; P. 189-200, 1968; P. 199-214, 1969.
1965: General review of technological developments, trade, production and planning in mineral industries. Detailed commodity review: iron and steel, other metals; asbestos, cement, fertilizer and chemical materials, fluorspar, magnesite, salt, other non-metals, coal, petroleum. Tabulated statistics: exports of selected metals and minerals to the USSR; imports of selected metals and minerals from the USSR; exports of selected metals and minerals to Japan; exports of selected metals and minerals to Poland.

1967: Mineral production affected by political turmoil associated with the Cultural Revolution. 'Every important sector of the mineral industry except salt, had a bad year.' Overall trade declined from 1966 levels, although minerals and metals remained significant in total trade; export of traditional commodities, import of fertilizers, steel products, some new plant, industrial and mining equipment (details). Commodity review of various metals, non-metals, mineral fuels. Tabulated statistics.
1968: Mineral output value in 1968 still had not returned to the $4-4.5 billion level estimated for 1966, although economic conditions did improve over 1967: PLA role in economic management, raw materials supply and products delivery systems brought under control by late 1968. Production, trade, commodity review: metals, non-metals, mineral fuels. 1969: Effect of political events on minerals production: 'The disastrous economic setback created by the Cultural Revolution was finally reversed . . . .' Petroleum spearheaded China's industrial recovery; coal industry 'went through a very peaceful and productive year': 'all the leading steel centers had good years.' Traditional export metals not available for sale in large quantities; imports of copper, lead, zinc, fertilizers. Review of trade. Commodity review of various metals, non-metals, mineral fuels.

625 WANG, K.P.
Mining and metallurgy. Sciences in Communist China, Gould. *AAAS* No. 68, P. 687-738, 1961.
General technological level of the larger mineral enterprises is just behind that in industrialized countries; some fields compare favourably. Mechanization lags behind basic engineering; thus small mines and plants operate beside large and efficient ones. Technological fields: mining; ore-dressing; metallurgy. Mineral industry sectors: coal; petroleum; iron and steel. Export metals: tin. Non-ferrous base and light metals. Non-metallics.

626 WANG, K.P.
Rich mineral resources spur Communist China's bid for industrial power. *Mineral Trade Notes*, Special Supplement No. 59. Washington, U.S. Dept. of the Interior, Bureau of Mines, 1960. Also Mineral wealth and industrial power. *Mining Engineering*, P. 901-12, Aug 1960. Also China's mineral wealth and metal production. *Journal of Metals*, P. 945-51, Dec 1960.
Survey of Chinese mineral resources, mineral industry and trade in minerals. Research and development in mineral industries (P. 15-16); level of mineral technology (P. 17-18). Contents: World significance of Chinese minerals; Growth of the mineral industry; Communist China's mineral self-sufficiency; Regional factors affecting mineral development; Management of the mineral industry; Present level of mineral technology; Major sectors of the mineral industry. The technological level of the Chinese mineral industry is uneven. The semi-modern operations are particularly suited for a country like China where equipment is inadequate and expensive and manpower abundant and relatively cheap. Generally speaking, the overall level of mineral technology in Communist China has advanced rapidly in the decade to 1960.

627 WANG, Kung-lee
China's mineral industries in 1967: victims of the Cultural Revolution. *Asian Survey*, No. 6, P. 425,37, Jun 1969.
Output of mineral and related commodities in 1967, based on 'little reliable data'; capital construction. Foreign trade: export of traditional Chinese metals, coal and salt, continuing import of fertilizer, steel products, increasing numbers of new plants; industrial and mining equipment. Iron and steel; petroleum, coal; non-ferrous metals.

628 WANG, Kung-lee
Mainland China. *Mining Annual Review* (published by *Mining Journal)*, P. 355-8, 1968. *(Mining Annual Review*, 1965, 1966, 1967, entry no. 615; 1969, entry no. 630; 1970, 1971, entry no. 623).

629 WEI, Anton W.T.
Minerals in China in 1961. *The Mining Journal*, Pt. I, P. 334-7, 6 Apr 1962; Pt. II, P. 359-62, 13 Apr 1962; Pt. III, P. 409-11, 27 Apr 1962.
Rapid development of mineral resources is the foundation for the emergence of an important industrial sector in China within a primarily agricultural economy. Economic setbacks in 1961 resulted in decreased labor- and plant-productivity; increased emphasis on quality, variety and safety braked production; 'withdrawal of Soviet technicians also will have a serious retarding effect on the future development of China's mineral industry.' Iron and steel; coal; petroleum; non-ferrous metals; ferro-alloy metals.

630 WU, Yuan-li
China. *Mining Annual Review* (published by *Mining Journal*), P. 381, 383, 1969.
'In general, and for the year as a whole, the output of mineral products was higher than in 1967, but still below that of 1966.' Improvement in transportation; iron and steel; coal; petroleum; foreign trade--in 1968. (*Mining Annual Review*, 1965, 1966, 1967, entry no. 615; 1968, entry no. 628; 1970, 1971, entry no. 623).

See also the following entries:

Technology Policy:
  Choice of Techniques. 40.
  Innovation and the Direction of Technological Change. 67, 91, 92, 97.
Policy for Science. 171, 173.
Industrial Technology:
  Electric Power Industry. 497.
  Petroleum Industry. 563.
Natural Resources. 650, 651, 652, 655, 656.

## C Engineering

631 AU, Lewis Li-tang
Civil and hydraulic engineering. Sciences in Communist China, Gould. *AAAS* No. 68, P. 771-804, 1961.
Civil engineering organizations; railroads and highways; bridges; buildings; structural analysis; hydraulic engineering. China has tried to follow Soviet theory and research in civil engineering and to catch up to western standards. Substitution of low- for high-cost materials. Western-trained engineers and scientists will have a more prominent role.

632 *CHINA NEWS ANALYSIS*
Engineering planning, No. 558, 2 Apr 1965.
'Revolution in engineering planning' in preparation for third FYP: survey of engineering capacity which takes the form of a political campaign. 'Engineering planning' drafting of all kinds, including architectural and engineering plans, the drawing of plans for buildings, factories, machines and modification of machines.

633 DEAN, Genevieve
A note on the sources of technological innovation in the People's Republic of China. *The Journal of Development Studies*, 9(1), 187-99, 1972.
The development of indigenous sources of technological innovation in the capital construction industries in China, with particular reference to the design reform movement of 1964-66. 'The main purpose of this note (is) to trace some of the changes in the organization of innovative activity in China which might account for the widening spectrum of technological options available to the planners in recent years' e.g. as inferred by Shigeru Ishikawa (entry no. 29).

634 NIEH, E.K.
Mechanical engineering. Sciences in Communist China, Gould. *AAAS* No. 68, P. 805-20, 1961.
Power generating equipment; boiler design and trend of development; machine building industry.

635 PAN, L.C.
Chemical engineering. Sciences in Communist China, Gould. *AAAS* No. 68, P. 821-40, 1961.
Survey of chemical engineering in China based on available publications; technical publications; reports from research laboratories; conferences and symposiums. Accomplishments in chemical engineering: synthetic ammonia; sulfuric acid; soda ash; chlorine-alkali industry; phosphate fertilizers; acetylene chemicals; petroleum refining; petrochemicals and high polymers. Miniature chemical plants. 'Chinese chemical engineers on the whole are making fair progress.'

636 SHERRARD, Howard
Notes on road and bridge engineering in China. *Eastern Horizon*,

7(5), 23-31, 1968.
Observations made during 1967 visit to China; administrative structure for highways in China; city and rural roads; structural features of specific bridges.

637 TSAO, T.C.
Electrical engineering. Sciences in Communist China, Gould. *AAAS* No. 68, P. 747-69, 1961.
Technological development given priority over basic research. Objectives set by 1956 National Industrial Construction Conference and Twelve Year Scientific Technology Plan. The economic framework. Foreign technical assistance. Power supply and manufacture of equipment; electrical communication and radio manufacture; electrical engineering research.

See also the following entries:

Technology Policy:
  Innovation and the Direction of Technological Change. 75, 84, 97, 99.
Policy for Science. 149, 200.
Scientific Communication. 378, 393.
Higher Education and Technical Training. 402, 412, 419.
Industrial Technology:
  Electronics Industry. 515.
  Chemical Industry. 525.
  Petroleum Industry. 551.
  Machinery Industry. 575.

## D Transportation and Communications

638 AKENO, Yoshio
The Chinese transportation industry. *Long-term prospects for the Chinese economy*, Pt. II, Chap. X, Shigeru Ishikawa (ed.). Tokyo, Asian Economics Research Institute, No. 102, 1966.

639 GOLDEN, Ronald
Peking pushes for an expanded international air transport system. *Aerospace International*, P. 18-22, May/Jun 1971.
Prospects of new Chinese international air services: China seeks to buy large aircraft abroad, especially in Britain. China produces versions of various Soviet aircraft. Development of Chinese electronics industry.

640 HUNTER, Holland
Transport in Soviet and Chinese development. *Economic development and cultural change*, 14(1), 71-84, 1965.
Lessons drawn from Soviet and Chinese experience include: 1--improved transport capacity is a concomitant of economic development, not a pre-condition for it; 2--a country's development priorities will strongly influence its choice of carriers . . . 'Communist China to date appears to be following faithfully in the Soviet pattern of carrier use.' '. . . revolutionary political leaders and their technical advisers, in both the USSR and China, have been exceedingly conservative in their transport technology.'

641 LIPPITT, Victor D.
Development of transportation in Communist China. *The China Quarterly*, No. 27, P. 101-19, Jul/Sep 1966.
Development of rail transport. Non-rail transport: shipping; motor vehicles.

642 ONOE, Etsuzo
Choice of industrial location in China. *Ajia Keizai*, 10(12), 27-46, 1969.

643 ONOE, Etsuzo
Regional distribution of industries in China: plan and idea. *Ajia Keizai*, 6(9), 78-89, 1965.

644 ONOYE, Etzuso
*Research on production location in China*. Tokyo, Institute of Developing Economies, 1971.
The question of choice of industrial location in China; changes in the policy of production location in new China; regional distribution of urban populations; deposits of resources and regional distribution of production; transportation of commodities; industry in Manchuria; industry in Shanghai; the iron and steel industry; energy production; cotton textile industry. Appendix: locational distribution of industry. Concerning the study of Yuan-li Wu (entry no. 647).

645 PORCH, Harriett E.
Civil aviation in Communist China since 1949. *RAND Memorandum* RM-4666-PR, Jun 1966.
Study of the development and operation of Mainland China's civil aviation since 1949; attempts to describe the air transportation system of Mainland China and to assess its role in serving the national economy. I--Development and organization of civil aviation in Communist China. II--Specialized aviation services. III--Air transport services. IV--Mainland China's civil aircraft and manufacturers.

646 VETTERLING, Philip W. and WAGY James J.
China: the transportation sector, 1950-71. *People's Republic of China: an economic assessment*, U.S. Congress, Joint Economic Committee. Washington, U.S. Govt. Printing Office, P. 147-81, 1972.
'Through a combination of modern and native transport facilities, the transportation sector of the PRC is solidly supporting Peking's industrialization drive.' Railroads; highways; inland waterways and coastal shipping; civil aviation. Tabulated statistics.

647 WU, Y.L., LING, H.C. and HSIA WU, Grace
*The spatial economy of Communist China. A study on industrial location and transportation.* New York, Frederick A. Praeger, 1967.
Every major economic decision necessarily entails an important locational decision--the significance of the facts of location in terms of economic development. Are the spatial arrangements of China's industrial activities co-ordinated with the national development program? Are spatial and transport decisions co-ordinated? See Etzuso Onoye, 'Research on production location in China' (entry no. 644).

See also the following entries:

Technology Policy:
  Choice of Techniques. 40.
  Innovation and the Direction of Technological Change. 67, 92.
Policy for Science. 159, 160, 171.
Research and Development:
  Military/Space Exploration and Travel/Nuclear Energy. 315.
Agricultural Technology. 491.
Industrial Technology:
  Electronics Industry. 507, 509.
  Petroleum Industry. 534, 536, 561.
  Mining Industry. 621.
Engineering. 636.

## E Natural resources

648 CHANG, K.S.
Geographical bases for industrial development in Northwestern China. *Economic Geography*, Vol. 39, P. 341-50, Oct 1963. Also *Annals of association of American geographers*, Vol. 52, P. 324, Sep 1962.
Brief note on the desirability of geographers' exploring the possibility for industrial development in the Northwest of China--study of distribution of various resources seems to indicate certain favorable conditions.

649 CHENG, Tien-hsi
Utilization of wild plants in Communist China. *Economic Botany*, 19(1), Jan/Mar 1965.
Wild plants as sources of edible and lubricating oils, starch, fibers, tannin, rubber, and as sources of raw materials for medicines, dyes, and insecticides. Mobilization of mass campaigns to identify and collect wild plants. 'Perhaps the most permanent contribution of the 'small autumn harvest' campaign is the revelation of the plant resources existing in China today.' However, the 'harvest' is a vivid example of lack of scientific thinking and planning on the part of the Communist rulers. Instead of allowing time for adequate survey, experimentation, installation of production facilities, and enactment of conservation laws, massive expeditions were suddenly organized by over-eager Party cadres to plunder the mountainous areas . . . . Owing to the lack of experimental data and technological know-how, finished products were generally inferior in quality.'

650 KARLSSON, Rolf
*China's energy resources: a literature survey*. Göteborg, Chalmers University of Technology, Oct 1971 (in Swedish).
Introduction; sources and evaluation; political and economic overview; population increase and agricultural production; China's geography and administration; China's foreign trade. Primary energy: coal; oil; natural gas; nuclear energy; hydropower; other fuels; production of primary energy in 1952, 1960, and 1970. Secondary energy: electricity; other secondary energy. Balance of energy. Future prospects. Conclusion.

651 NIU, Sien-chong
China's mineral wealth. *Ordnance*, Vol. 51, P. 587-90, May/Jun 1967.
'China probably possesses very considerable mineral resources,' now being explored by 4000 prospecting and surveying teams and 21000 geological experts. Survey of coal, iron ore, oil, uranium, copper molybdenum, tungsten and aluminium resources, drawing on Y.M. Berger, *Kitai; ekonomiko-geografischeskii-ocherk* (1959).

652 *NOTES ET ETUDES DOCUMENTAIRES*
Les sources d'energie en Republique Populaire de Chine, No. 2804, 12 Aug 1961. Paris, Secretariat General du Government, Direction de la Documentation.

Coal: general remarks, history, basins. Hydroelectric power: utilization of resources for hydroelectric power, development of the electric industry. Petroleum: bituminous schists, natural gas, prospecting, distribution of strata.

653 *SCIENCE NEWS LETTER*
Minerals found in China, Vol. 79, P. 3-4, Jan 1961.
Brief note based on interview with Dr. Edward C.T. Chao: 'Recent valuable mineral discoveries in Red China are due to the vast increase in the number of her trained scientists since the Communist regime came to power.'

654 WANG, K.P.
The mineral resource base of Communist China. *An economic profile of Mainland China,* U.S. Congress Joint Economic Committee. New York, Praeger, P. 167-95, 1968.

655 *WORLD OIL*
What's happening around the world: China, 143(3), 381-2, 1956.
Brief review of China's potential as an oil producer. Geological structures 'indicate that prospects are good for development of large oil and gas reserves.' 'The remoteness of the lands, and lack of good roads, technical personnel, adequate equipment and good communications are the only apparent hindrances to China's oil expansion.'

656 WU, Yuan-li with LING H.C.
*Economic development and the use of energy resources in Communist China.* New York, Praeger, for the Hoover Institution, 1963.
'This study of the use of energy resources in Communist China is essentially an inquiry into a particular aspect of the process of economic growth.' Conclusions: existence of adequate to abundant reserves; growing contribution of energy resources to national income; high industrial consumption of electricity; low electricity consumption per industrial worker; coal the most important source of energy; industry the principal consumer of energy products; development of energy resources in response to expected demand from industrial and other planned developments. Chap. 2--The growth of the electric power industry, 1949-60. 3--The growth of coal mining, 1949-60. 4--The supply and demand of electric power. 5--The supply and demand of coal. 6--Sales, costs, and contribution to national income. 7--The optimal use of energy resources in national and regional development. 8--The infant petroleum industry.

See also the following entries:

Technology Policy:
  Technology and Economic Growth. 4.
  Choice of Techniques. 40.
  Innovation and the Direction of Technological Change. 97.
Policy for Science. 121, 171.
Agricultural Technology. 476, 479.

Industrial Technology:
Petroleum Industry. 533, 534, 536, 539, 540, 542, 543, 544, 551, 559, 560, 562.
Basic Metals Industries. 601, 610.
Mining Industry. 615, 616, 618, 619, 622, 624, 626, 627, 628, 629.
Transportation and Communication. 644, 647.

# F Trade

657 CHAMBERS, Sir Paul
Trading with China. *Far East trade and development*, 21(2), 125-7, 1966.
Chairman of Imperial Chemical Industries reports on ICI trade with China and its exhibition in Tientsin.

658 DALYELL, Tam
China--a market for British know-how now. *New Scientist*, 52(775), 222-4, 1971.
Description of the author's visit in 1971 to Canton trade fair, Shanghai Industrial Exhibition, Capital Iron and Steel Plant, machine tool factories in Peking and Shanghai, a Shanghai truck plant, Peking nos. 1, 2 and 3 Cotton Factories, Shanghai Chemical Plant. Prospects for trade with China: not an easy market, but a definite role exists for smaller British firms if they act 'now, as China opens up to the world.'

659 DALYELL, Tam
Scottish mission to China stresses stamina to sustain technical cooperation. *Far East trade and development*, 27(1), 33-4, 1972.
Description of a trip to China by trade mission of the Scottish Council for Development and Industry, Nov 1971: Kwangchow Autumn Commodities Fair, various Peking and Shanghai factories, Shanghai Industrial Exhibit. The mission was 'most impressed' by machine tools produced in China. 'The deduction of the delegation was that there is little the Chinese cannot do in designing highly technical machines for production in small numbers.' Interest expressed by the Chinese in prospects for exporting simple, easily-maintained equipment to developing countries; determination to develop China without foreign dependence--'In the eyes of the Chinese government, trade with the outside world is mainly about the speed of industrial advance . . .'

660 *FAR EAST TRADE AND DEVELOPMENT*
Towards positive trade relations, 27(10), 435-8, 1972.
Review of China's policies governing foreign trade, as enunciated by the Chinese delegation to UNCTAD III: technology transfer; balanced trade; international monetary reform. Pattern of trade. Kwangchow Trade Fair. Chinese aid. P. 442-3: Trading Corporations of China.

661 *FAR EASTERN ECONOMIC REVIEW*
Deals without dependence, 74(40), 34, 36-7, 1971.
Foreign trade. Production for export. Imitation and modification of foreign techniques.

662 *FAR EASTERN ECONOMIC REVIEW*
Tokyo-Peking trade winds, 49(9), 378-80, 1965.
Japanese sales of plant to China.

663 KESWICK, John
The British Exhibition in Peking, Nov 1964. *Journal of the Royal*

*Central Asian Society*, Vol. 52, P. 275-80, Jul/Oct 1965.
Address to the Royal Central Asian Society, 22 Feb 1965. Peking Industrial and Machinery Exhibition sponsored by the Sino-British Trade Council. British exports to China.

664 KIRBY, E. Stuart
Trade and development of Mainland China. *Contemporary China: papers presented at the University of Guelph Conference, April 1968*. Toronto, The Canadian Institute of International Affairs, P. 62-97, 1968.
Survey of the patterns of China's international trade. The developing countries, and particularly China, are intent on industrializing; China must rely on foreign trade for modern capital goods, technologically advanced and sophisticated equipment.

665 MACDOUGALL, Colina
Eight plants for Peking. *Far Eastern Economic Review*, 43(4), 155-8, 1964.
China's trade with the West. Purchase of complete plant.

666 OLDHAM, C.H.G.
The challenge of China. *The Business Quarterly*, 32(2), 48-50, 1967.
Discussion of possible development of Canadian relations with China. Foreign trade; China needs to import advanced technology in the medium term.

667 OLDHAM, C.H.G.
*The Chinese Trade Exhibition in Japan*. Institute of Current World Affairs, unpublished newsletter CHGO-28, 18 Jul 1964.
Description of Chinese trade exhibition, including machine tools and geophysical instruments.

668 WILSON, Dick
China's trading prospects. *Far Eastern Economic Review*, 45(8), 309-11, 1964.
China imports materials and equipment for the chemical, oil and textile industries.

669 WILSON, Dick
Peking's trading plans. *Far Eastern Economic Review*, 48(8), 352-4, 1965.
China's foreign trade. Orders for complete plant.

670 WILSON, Dick
Showing off in China. *Far Eastern Economic Review*, 44(4), 197-8, 1964.
Western trade and industrial exhibitions in China. SIMA exhibition.

See also the following entries:

Technology and Economic Growth. 4, 10, 13, 26.
Technology Policy:
Choice of Techniques. 36.
Transfer of Technology. 44, 48.

Industrial Technology:
Electronics Industry. 513, 517.
Chemical Industry. 525, 532.
Petroleum Industry. 536, 539, 540, 542, 543, 551, 557, 558, 560, 567, 569.
Machinery Industry. 576.
Mining Industry. 623, 624, 625, 626.

# G Technology and Employment

671 CHAO, Kuo-chun
Rural manpower in India and China. *Far Eastern Economic Review*, 35(2), 53-63, 1962.
Comparison of India's utilization of rural manpower with similar problems in China. Shortage of labor power developed in many rural areas of China since 1957. Effective organization and fuller exploitation of potential resources would ease the problem of under-utilization of rural manpower in India.

672 CHEN, Pi-chao
Over-urbanization, rustication of urban-educated youths, and politics of rural transformation. *Comparative Politics*, P. 361-86, Apr 1972.
'. . . most under-developed countries have failed to attract and put to work in the rural areas those persons who have received modern education and cannot find jobs in the urban areas. This article purports to describe and analyze the attempt of one under-developed country, China, to solve this contradiction by resettling urban school graduates in the countryside and employing them there as catalytic agents of rural development.' Pt. IV--The economic calculus of stabilizing urban population growth and rustication. The dualist approach to labor absorption in rural industries and agriculture, technological modernization in urban industries.

673 *CHINA NEWS ANALYSIS*
Manpower distribution and migration, No. 95, Aug 1955.
Skilled workers, P. 2. University graduates, P. 5.

674 *CHINA NEWS ANALYSIS*
Manpower survey, No. 169, Feb 1957.
Based on *Bulletin for statistical work*, 'Number, structure and distribution of workers and staffs in 1955'. Including redundancy in administration, skill (by age), regional distribution.

675 ECKSTEIN, Alexander
Manpower and industrialization in Communist China, 1952-1957. In 'Population trends in Eastern Europe, the USSR and Mainland China'. *Proceedings of the 36th Annual Conference of the Milbank Memorial Fund, Nov 1959*. New York, Milbank Memorial Fund, P. 157-68, Discussion P. 168-176, 1960.
Differences in the patterns of agrarian transformation in Russia and China as they affect the inter-relationships between manpower movements and industrialization.

676 EMERSON, John Philip
Manpower absorption in the non-agricultural branches of the economy of Communist China, 1953-1958. *The China Quarterly*, No. 7, P. 69-84, Jul/Sep 1961.
In 1968 the previous trend of urban population growing faster than non-agricultural or urban employment was reversed; non-agricultural and urban employment grew so much that a migration from countryside

to town of unprecedented magnitude occurred. The direction of this expansion of labor inputs was towards industry, construction and modern transport.

677 ISHIKAWA, Shigeru
Change of employment and structure of productivity in Communist China. *Ajia Keizai*, 2(1), 2-14, 1961.

678 KRADER, Lawrence and AIRD, John
Sources of demographic data on Mainland China. *American Sociological Review*, 24(5), 623-30, 1959.
A radical shift was made in the official policy on birth control in response to what was discovered about population size and rate of growth . . . . Urban labor force was growing more rapidly than capacity of urban industry to absorb it: consequently current policy discourages migration of rural residents to the cities. These major policy decisions testify to the acceptance of official population data.

679 ORLEANS, Leo A.
Population redistribution in Communist China. In 'Population trends in Eastern Europe, the USSR and Mainland China'. *Proceedings of the 36th annual conference of the Millbank Memorial Fund, Nov 1959*. New York, Millbank Memorial Fund, P. 141-50, Discussion P. 150-56, 1960.
Relative stability in the distribution of China's population; highly unbalanced density pattern. Migration from rural to urban areas, 1949-1956. Development of industrial bases in Western China. Resettlement of population on reclaimed agricultural lands. Spatial redistribution of the Chinese population, 1949-1959, has been relatively slight.

680 ORLEANS, Leo A.
Problems of manpower absorption in rural China. *The China Quarterly*, No. 7, P. 57-68, Jul/Sep 1961.
A real and serious labor shortage has been evident in agricultural production, but this is due to mass labor projects and inefficient utilization of available manpower.

681 ORLEANS, Leo A.
The recent growth of China's urban population. *The Geographical Review*, 49(1), 43-57, 1959.
Size and rate of growth of China's urban population to some extent reflects the industrialization of China. Conclusion: the balance between urban and rural populations will remain virtually constant for several decades, despite an accelerated program of industrialization.

682 TIEN, H. Yuan
Educational expansion, deployment of educated personnel and economic development in China. *World population conference, Belgrade, 1965: proceedings*, Vol. 4. P. 181-3. New York, United Nations, 1967.
'. . . the economic development of China thus far has allowed only a limited change in occupational structure . . . the expansion in

education, especially at the primary and secondary levels, has evidently occurred at a much faster rate than employment opportunities in the non-agricultural sector.'

See also the following entries:

Technology and Economic Growth. 4, 9, 10.
Technology Policy:
  Choice of Techniques. 35.
  Innovation and the Direction of Technological Change. 94, 106, 107.
Higher Education and Technical Training. 404, 412, 419, 433.
Transportation and Communications. 644, 647.

# H Medicine and Public Health

683 AGREN, Hans
Medical practice in China: a compendium. *Science,* 178(4059), 394-5, 1972.
Book review of *Nongcun Yisheng Shouce (Peasant village physician's handbook),* compiled by the Medical Revolutionary Committee of Hunan. 'The present book . . . is intended for general use in the Chinese countryside, to be studied as a textbook by all those having to deal with health- and sick-care and used as a handbook by doctors with more extensive theoretical training.'

684 ALLEN, Edwin J. Jr.
*Disease control in China.* Columbia University, M.A. Dissertation, 1965.
'An investigation into the ways in which public health propoganda effects changes in medicine and hygiene, with emphasis on schistosomiasis control.'

685 *AMERICAN MEDICAL NEWS*
Inside look at Chinese medicine, Vol. 14, P. 1, 10, Oct 1971.
Report of a visit to China by Dr. E. Grey Dimond in Sept 1971. Observations on hospital equipment, manufacture of equipment and drugs in China. Combination of Western and Chinese medicine. Medical research concentrates on applied research, medical education shortened. No national medical publications, although medical journals are received in China from the US. No travel abroad by Chinese medical personnel 'until the goals of the Cultural Revolution are met.' Acupuncture anesthesia. Family planning and birth control.

686 BAIRD, John
Public health in China. *United Asia,* 8(2), 114-16, 1956.
Medical institutes: Shenyang Medical Institute. Teaching; clinical and pathological research emphasized over fundamental research; reassessment of traditional Chinese medicine. Szechwan Medical Institute.

687 BEST, J.B.
Impressions of Chinese medical services. *Eastern Horizon,* 6(10), 32-5, 1967.
Observations of 1966 visit to China by Australian physician. Western and Chinese medicine; Ministry of Public Health; discussion with Director of Peking Institute of Dermatology and Venerology; medical education; medical supplies.

688 BOWERS, John Z.
Medicine in Mainland China: Red and rural. *Current Scene,* 8(12), 1970. Also in *Population and family planning in the People's Republic of China,* Piotrow, Phyllis T. (ed.). Washington, The Victor-Bostrom Fund and the Population Crisis Committees, P. 22-4, 1971.
Traditional medicine; history of the effort to fuse Chinese and Western medicine. Barefoot doctors; new approaches to acupuncture. Medical education. Family planning. Health services.

689 CHEN, Paul
Acupuncture and moxibustion. *Eastern Horizon,* 5(6), 29-33, 1966.

690 CHEN, William Y.
Medicine and public health. In 'Sciences in Communist China', Gould. *AAAS* No. 68, P. 383-408, 1961. Also in *The China Quarterly,* No. 6, P. 153-169, Apr/Jun 1961.
Medical education; emphasis on quantity, improvement in quality; integration of traditional and modern medicine; practical field education; publication. Progress in medicine. Traditional medicine. Public health; preventive medicine and sanitation; medical care. Central Research Institute of health. Cancer prevention and control.

691 CHENG, Huan
There is something to it. *Far Eastern Economic Review,* 75(12), 22-3, 1972.

692 CHENG, Tien-hsi
Disease control and prevention in China. In 'Science and medicine in the People's Republic of China'. *Asia,* No. 26, P. 31-59, 1972. 'In the last two decades, numerous innovations and changes have been instituted in China for disease control and prevention. Included in this report are highlights of some major programs dealing with parasitic diseases, cancer and mental disorders.'

693 CHENG, Tien-hsi
Schistosomiasis in Mainland China. *The American Journal of Tropical Medicine and hygiene,* 20(1), 26-53, 1971.
'This report is based on a review of about 250 published articles, mostly technical; on interviews with Chinese expatriates, from the mainland, now residing in Hong Kong and Macao, and with Japanese experts who have visited China; and on surveys of commentaries and news releases from communist, neutral, and anti-communist sources. The contents relate to highlights of anti-schistosomiasis activities with emphasis on epidemiology, prevention, and treatment of the disease. Regarding morphology, taxonomy, and biology of the parasite or the vector snail, we mentioned only findings heretofore not cited or published outside mainland China. In the concluding paragraphs, considerable space is devoted to economic, social, and political impacts of the antischistosomiasis campaign, which may serve as a timely reference for developing countries engaged in the same struggle.'

694 *CHINA NEWS ANALYSIS*
Health, No. 365, Mar 1961.
Evidence of epidemics; causes and prevention.

695 *CHINA NEWS ANALYSIS*
Medicine in China, No. 98, Sep 1955.
Hospital and doctors. Medical education. Russian medicine.

696 *CHINA NEWS ANALYSIS*
Towards a new science of medicine? No. 269, Mar 1959.
The two medicines (Western and traditional). Epidemics. Drugs.

697 CHINA RESEARCH INSTITUTE (Tokyo)
Medicine and hygiene in New China. *China Materials Monthly Bulletin*, No. 55, Aug 1952.
Medicine and hygiene of the past and its overthrow; Hygiene work among the people; Preventive and environment hygiene; Herbal medicine; Germ warfare and the patriotic health campaign; People's physical education.

698 CROIZIER, Ralph C.
Chinese Communist attitudes toward traditional medicine. *Asia*, Vol. 5, P. 70-76, Spring 1966. Also in *China's cultural legacy and Communism*, Croizier, Ralph C. (ed.). London, Pall Mall Press, P. 270-5, 1970.
Fluctuation in official policy toward traditional Chinese medical practice. National pride and cultural diplomacy, practical factors as motivations for emphasizing Chinese medicine. 'It is precisely because so much of China's total social, political, and cultural situation has changed that Mao and company . . . have been able to cast Chinese medicine in a new positive role in the new China.

699 CROIZIER, Ralph C.
Traditional medicine in Communist China; science, Communism and cultural nationalism. *The China Quarterly*, No. 23, P. 1-27, Jul/Aug 1965.
History of Chinese Communist attitude toward traditional medicine from the 1930s to after the Great Leap Forward.

700 CROIZIER, Ralph C.
*Traditional medicine in Modern China: science, nationalism, and the tensions of cultural change*. Cambridge, Harvard University Press, 1968.
'The central paradox and main theme is why twentieth century intellectuals, committed in so many ways to science and modernity, have insisted on upholding China's ancient prescientific medical tradition' . . . . 'With science generally hailed as the key to national suvival, and almost all that science of foreign origin, claims for the scientific validity of the native medical tradition have had a powerful appeal to cultural nationalism.' The relationship of medicine to cultural and intellectual developments; 'medical acculturation.' Pt. 1--Traditional Chinese medicine and the introduction of modern medicine. Pt. 2--Medicine as a cultural and intellectual issue, 1895-1949. Pt. 3--Science, Communism, and 'the People's medical heritage.

701 *CURRENT SCENE* EDITOR
Mao's revolution in public health. *Current Scene*, 6(7), 1968.

702 DIMOND, E. Grey
Medical education and care in People's Republic of China. *The Journal of the American Medical Association*, 218(10), 1552-1557, 1971.
'What the Chinese are doing and trying to do in medical education and medical care can only be understood in terms of the population distribution of China and of the consequences of the second or Cultural Revolution.' 'The changes in medical education and medical

care can best be described as a change in priorities': medical care must be available now to all the people; education must be controlled by the people, not by an academic framework as we know it. Synopsis of events from 1966 to the present in medical education. Barefoot doctors. Description of a commune hospital. Research at the Institute of Materia Medica, CAS.

703 DURDIN, Peggy
Medicine in China: a revealing story. *New York Times Magazine*, P. 17, 76-9, 28 Feb 1960.
Research on and use of traditional Chinese medicine.

704 ELLIOTT, K.A.C.
Observations on medical science and education in the People's Republic of China. *The Canadian Medical Association Journal*, Vol. 92, P. 73-6, Jan 1965.
Observations from the author's trip to China in 1964. Table: organization of medical science and teaching, Peking and Shanghai. Basic research in Chinese Medical College. Peking Medical College. The Chinese Academy of Medical Sciences. The Peking Union Hospital. Bethune Peace Hospital. Chung Shan Medical College. Shanghai First and Second Medical Colleges; Railway Medical College. 'The headquarters of the Chinese Academy of Sciences . . . are in Shanghai . . . . Its director is Dr. T.P. Feng.' Institute of Physiology. Preventive medicine and medical service to rural areas are especially fostered. Research programs throughout the country are based on plans of the central, provincial and local governments, and of the particular institute, and on the preferences of the individual worker.

705 ESPOSITO, Bruce J.
The People's Liberation Army, medicine and the Cultural Revolution. *Marine Corps Gazette*, Jun 1971.
'The PLA has immeasurably aided the development and proliferation of available medical services in the countryside.'

706 ESPOSITO, Bruce J.
The politics of medicine in the People's Republic of China. *Bulletin of the Atomic Scientists*, 28(10), Dec 1972.
Health care policies in China during the Cultural Revolution. Medical assistance in rural areas will 'definitely improve', but there will probably be 'a reduction in the health standard of urban areas.' Regular medical service in the PLA has been overextended. Disruption of medical schools since 1966 will reduce the number of medical specialists and Western-trained doctors and 'will certainly have detrimental effects on the general health of the population.'

707 GALSTON, Arthur W.
Attitudes on acupuncture. *Natural History*, Vol. 81, P. 14-16, 92, 1972.
Controversy in Western medical circles following the report by the author and Ethan Signer of their observations of acupuncture anesthesia in China in 1971. 'This lack of an adequate body of theory to account satisfactorily for the efficacy of the described

(acupuncture) points has led, inevitably, to a rejection by Western doctors of the alleged efficacy itself.' The use of electricity in conjunction with acupuncture was introduced during the Cultural Revolution; but was only reported in 'internal information bulletins circulating among Chinese medical facilities.'

708 HAN, Suyin
Acupuncture--the scientific evidence. *Eastern Horizon*, 3(4), 8-17, 1964.

709 HIEDA, Kentaro
Concerning medical science in China. *Ajia Keizai*, 11(9), 2-25, 1970.

710 HORN, J.S.
*Away with all pests . . . an English surgeon in People's China*. London, Paul Hamlyn, 1969.
Experiences of an English surgeon in China in 1937 and from 1954 onwards. Account of 'human relationships in the hospital,' campaigns against syphilis and schistosomiasis, health measures in rural areas. 'In this book I have described some of New China's achievements in the field of medicine . . . I am confident that as a result of the Cultural Revolution, and achievements of the coming decades will dwarf those which I had the privilege to witness.'

711 HORN, J.S.
Quantity and quality in surgery. *Arts and Sciences in China*, 1(4), 20-22, 1963.
Two methods to promote quality of surgery: emulation of outstanding medical successes; surgical congress.

712 HORN, J.S.
Breakthrough tactics in Chinese surgery. *Eastern Horizon*, 3(1), 32-4, 1964.
Rapid quantitative expansion of Chinese medical services after 1949. Quality upgraded by the 'breakthrough' method.

713 HU, Chang-tu, *et al*
*China: Its people, Its society, Its culture*. New Haven, HRAF Press, 1960.
Chap. 19--Public health and welfare. 20--Education. 21--Science and technology. Selected bibliography.

714 HUARD, Pierre and WONG Ming
*Chinese Medicine*, trans. Bernard Fielding. London, Weidenfeld and Nicolson Ltd., 1968.
'. . . for want of the necessary substructure, (Western medicine) has not yet been able to drive out popular medicine which . . . is admirably suited to the economic and psychological condition of the peasant masses who support it . . . . This traditional medicine . . . is of interest not only to doctors through its extremely varied therapeutics, but also to linguists, ethnologists, sociologists, psychologists and all who are aware of the fruitful possibilities of a medically informed approach to the study of

China and its potentialities for modern science.' Historical review of the evolution of Chinese medicine (in the Ancient Empire, 206 BC - AD 580; classical medicine, Sui and T'ang; modern medicine, Ming and Ch'ing). Chinese medicine, the medicine of other Asiatic countries, and European medicine, Western medicine in modern China, including medical education, the pharmaceutical industry. Traditional medicine in modern China. References.

715 LACOUTURE, Jean
La 'Medecine Mao Tse-toung'. *La Recherche,* 3(23), 421-2, 1972. Reorganization of Shanghai No. 6 Hospital as a result of the Cultural Revolution. Synthesis of traditional medicine and modern science.

716 LEE, Tsung-ying
Synthesizing Chinese and Western medicine. *Eastern Horizon,* 8(6), 33-7, 1969.
Development of Chinese medicine is not a stop-gap measure to supplement inadequate modern medical facilities, but is being developed together with Western medicine until 'both are eventually merged into a new medical science much more highly developed than either of the two.'

717 MADDIN, Stuart
Medicine in China today. *Eastern Horizon,* 11(6), 31-6, 1972. Canadian dermatologist describes his 1972 visit to China. Medical education; formulation of medical policy by revolutionary committees. Extension of modern medical facilities to the countryside. Acupuncture anesthesia; theory of acupuncture. Herbal medicine and research on traditional herbal medicines. Barefoot doctors. Commune cooperative medical associations.

718 MAEGRAITH, Brian
The Chinese are 'liquidating' their disease problems. *New Scientist,* 3(55), 19-21, 1957.
Report by a British scientist 'recently returned from a visit to China' on policies to eliminate public health hazards: campaigns to promote hygiene, eliminate schistosomiasis, gastro-intestinal diseases, malaria.

719 MANN, Felix
Chinese traditional medicine: a practitioner's view. *The China Quarterly,* No. 23, P. 28-36, Jul/Aug 1965.
Observations from a visit to China in 1963 of Chinese medical practice. Descriptions of acupuncture.

720 *MD MEDICAL NEWSMAGAZINE*
China, 15(2), 115-30, 1971.
General survey of Chinese history. History of Chinese medicine, p. 127-9. Medicine in modern China, p. 130.

721 PENFIELD, Wilder
Oriental renaissance in education and medicine. *Science,* 141(3586), 1153-1161, 1963.

Observations of a 1962 visit to China by a Canadian physician. Medical education; traditional medicine; traditional and modern surgery; common diseases; population and birth control.

722 RIFKIN, Susan B.
Doctors in the fields. *Far Eastern Economic Review*, 75(12), 20-22, 1972.

723 RIFKIN, Susan B.
Health services in China. *Bulletin of the Institute of Development Studies*, 4(2/3), 32-8, 1972.
Extension of health services to the rural population.

724 SALAFF, Janet
Physician heal thyself. *Far Eastern Economic Review*, 62(44), 291-3, 1968.
Medicine and public health in the Cultural Revolution.

725 SIDEL, Victor W.
Serve the people: medical care in the People's Republic of China in 'Science and Medicine in the People's Republic of China'. *Asia*, No. 26, P. 3-30, Summer 1972.
Text of a talk at The Asia Society in Spring, 1972. Lane health stations and 'red guard doctors' delivery of neighborhood health care. Birth control and abortion. Integration of traditional and modern Western-type medicine: acupuncture anesthesia. Commune health stations. Production brigade health stations; barefoot doctors. Industrial medicine and occupational health. Sun Yat-sen Medical School.

726 SIEH, Marie
Medicine in China: wealth for the state. *Current Scene*, Vol. 3, No. 5, Pt. I, 15 Oct 1964; No. 6, Pt. II, 1 Nov 1964.
Description of medical care and training in China based on interviews with refugees.

727 STOVICKOVA, Dana
What is acupuncture? *Eastern Horizon*, 1(8), 11-18, 1961.

728 TAKAHASHI, KEIJI, Yamada, ICHII, Saburo
Theory and practice as seen in the School of Chinese Medicine. *Chugoku*, No. 88, P. 6-28, Mar 1971.

729 THOMSON, R.K.C., MACKENZIE, Walter C., PEART, A.F.W.
A visit to the People's Republic of China. *The Canadian Medical Association Journal*, 97(7), 349-60, 1967.
Account of a visit to China in November 1966. 'The object of the visit to China was to observe medical education, medical research and medical practice in China . . . .' Description of visits to . . . The Institute of Mental Health, The Institute of Heart Disease, The Institute of Industrial Health and Environmental and Nutritional Health. The Research Institute of Acupuncture and Moxabustion; The Institute of Vaccines and Biological Products. Sun Yat-sen Medical College.

730 UNGER, Jonathan
On snails and pills. *Far Eastern Economic Review*, 74(40), 24-6, 1971.
Health services, public health, contraception.

731 WILLOX, G.L.
Contemporary Chinese health, medical practice and philosophy in *Contemporary China*, Ruth Adams (ed.). New York, Vintage Books, P. 105-120, 1966. Also in *Bulletin of the Atomic Scientists*, 22(6), 51-6, 1966.
Observations by a Canadian doctor of a visit to China in 1964. Medicine and medical training; research facilities; salaries; Norman Bethune; hospital visits; traditional Chinese medicine; communes and local medical facilities; surgical practice; conditions in Shanghai; birth control; Christianity.

732 WORTH, Robert M.
Health trends in China since the Great Leap Forward. *The China Quarterly*, No. 22, P. 181-9, Apr/Jun 1965.
Observations based on examination of children recently arrived in Macao from China and comparison with Hong Kong children; and on interview with nine doctors who left China in 1962.

733 WORTH, Robert M.
Strategy of change in the People's Republic of China--the rural health center in *Communication and change in the developing countries*, Lerner, David and Wilbur Schramm, (eds.). Honolulu, East-West Center Press, P. 216-30, 1967. Also Institution Building in the People's Republic of China: the rural health center. *East-West Center Review*, 1(3), 19-34, 1965.
'The authorities in Mainland China have attacked their health problems with a creative blend of well-demonstrated public health techniques plus some novel innovations that have maximised the usefulness of their human resources and at the same time have satisfied certain political and psychological needs.' Rural health centers provide a technical service to rural people, alleviate shortage of medical manpower by utilizing traditional practitioners, identify scientifically valuable aspects of Chinese medical tradition.

See also the following entries:

Policy for Science. 150, 167, 171, 190, 200.
Research and Development:
Military/Space Exploration and Travel/Nuclear Energy. 359.
Scientific Communication. 394.
Higher Education and Technical Training. 413, 417, 419, 431, 440, 454.
Population Control. 744.
Environmental Control. 756.

# I Population Control

734 AIRD, John S.
Population policy and demographic prospects in the People's Republic of China in *People's Republic of China: an economic assessment*, U.S. Congress, Joint Economic Committee. Washington, U.S. Govt. Printing Office, P. 220-331, 1972.
P. 253-69: Methods of contraception.

735 CHEN, Pi-chao
China's birth control action programme, 1956-1964. *Population Studies*, P. 141-58, Jul 1970.
I--The organizational set-up. II--Communication and persuasion. III--Means and services provided: techniques and the development of techniques for birth control. IV--Delayed marrriage. V--Strategy of implementing action programmes.

736 *CHINA NEWS ANALYSIS*
Planned birth rate, No. 172, 15 Mar 1957.

737 EHRLICH, Paul R. and HOLDREN, John P.
Neither Marx nor Malthus. *Saturday Review*, 54(45), 88, 1971.
'Family planning services . . . have been available in China for many years, hindered only by limitations of production, distribution, and peasant education . . . . In addition, however, population control measures . . . have been pushed vigorously in China, if somewhat intermittently.' Population control campaigns in 1956, between 1962 and 1966, and since 1968. Techniques of birth control: 'an apparently sophisticated contraceptive pill'; abortion by the vacuum method; extension of 'decent' medical care, contraceptive services, and effective birth control propaganda by paramedical barefoot doctors; equality of Chinese women with men.

738 FAUNDES, Anibal and LUUKKAINEN, Tapani
Health and family planning services in the Chinese People's Republic. *Studies in family planning*, 3(7) (supplement), 165-76, 1972.
Travel report by Chilean and Finnish obstetrician/gynecologists following visit to China, March 1972. Observations on health care service, organization and delivery of general health services, organization and delivery of maternal health care and family planning services, family planning methods. Research on and development of techniques of population control in China.

739 FREEBERNE, Michael
Birth control in China. *Population Studies*, 18(1), 5-16, 1964.
Birth control campaign, 1956-1958; 1962-1963; campaign against early marriages. Improved standards of education and communication facilitate dissemination of knowledge about birth control, but no breakthrough in population control has yet been achieved; traditional methods are being utilized.

740 FREEBERNE, Michael
The spectre of Malthus: birth control in Communist China. *Current Scene*, 2(18), Aug 1963.
Birth control campaigns in China.

741 HAN, Suyin
Family planning in China. *Japan Quarterly*, 17(4), 433-42, 1970.
Also *La Nouvelle Chine*, No. 1, P. 37-41, Mar 1971. Also *Population and family planning in the People's Republic of China*, Piotrow, Phyllis T. (ed.). Washington, The Victor-Bostrom Fund and the Population Crisis Committees, P. 16-17, 19-20, Spring 1971.
Present rate of population growth in China is 2% per annum; there is no 'population explosion' in China. Planned parenthood campaigns began in China in 1956 in the cities; with the Sociolist Education movement, family planning took root in rural areas (1963). Use of mobile medical teams in rural areas. The basis of family planning in China is the emancipation of women, not economic coercion. Political struggle over health and medicine policies on the eve of the Cultural Revolution (1965). Barefoot doctors; family planning now involves mass participation, knowledge and decision-making. Techniques of birth control (sterilization, mechanical techniques, oral contraception); abortion; correction of infertility.

742 HAN, Suyin
Family planning in China today. *Eastern Horizon*, 4(11), 5-8, 1965.

743 ORLEANS, Leo A.
Birth control: reversal or postponement? *The China Quarterly*, No. 3, P. 59-73, Jul/Sep 1960.
Birth control campaigns will be limited until the means are available in quantity and results can be assured.

744 ORLEANS, Leo A.
China's population: reflections and speculations in *Contemporary China*, Ruth Adams (ed.). New York, Vintage Books, P. 239-52, 1966.
Also *Bulletin of the Atomic Scientists*, 22(6), 22-6, 1966.
There is some evidence that the birth control campaign now in progress has considerable impetus and is operating through the public health system, but no indication that the regime has assigned to it the high priority necessary for effective fertility control.

745 ORLEANS, Leo A.
Evidence from Chinese medical journals on current population policy. *The China Quarterly*, No. 40, P. 137-46, Oct/Dec 1969.
In the 1960s the effort to limit the birth rate was directed at professional medical and public health personnel through the more sophisticated medical journals.

746 ORLEANS, Leo A.
A new birth control campaign? *The China Quarterly*, No. 12, P. 207-10, Oct/Dec 1962.
Implementation of birth control policies conflicts with ideology.

747 ORLEANS, Leo A.
*What is new in birth control in China,* unpublished draft n.d.
Brief review of Chinese Communist birth control policies; dissemination of birth control information; abortion; sterilization; contraception. 'China has reached a point at which fertility control is causing a downward trend in the country's birth rate.'

748 PIOTROW, Phyllis T. (ed.)
*Population and family planning in the People's Republic of China.* Washington, The Victor-Bostrom Fund Committee and the Population Crisis Committee, Spring 1971.
Contents: Edgar Snow, Population care and control; Han Suyin, Family planning in China; John Z. Bowers, China's Medicine: Red and rural; Huang Yu-chuan, Birth control education campaigns.

749 TIEN, H. Yuan
Population control: recent developments in Mainland China. *Asian Survey,* 2(5), 12-16, 1962.

750 TIEN, H. Yuan
Sterilization, oral contraception, and population control in China. *Population Studies,* 18(3), 215-35, 1965.

See also the following entries:

Policy for Science. 207.
Higher Education and Technical Training. 431.
Medicine and Public Health. 685, 688, 721, 725, 730, 731.

## J Environmental Control

751 CARIN, Robert
Irrigation scheme in Communist China. *Communist China Problem Research Series EC33*. Hong Kong, Union Research Institute, 1963.
Analysis of the political and economic-geographic changes resulting from Chinese Communist irrigation schemes.

752 CARIN, Robert
River control in Communist China. *Communist China Problem Research Series EC31*. Hong Kong, Union Research Institute, 1962.
Description of flood control projects. 'In spite of all attempts to control rivers, China's flood problem remains unsolved.'

753 DAWSON, Owen L.
Irrigation developments under the Communist regime. *Food and agriculture in Communist China*, Buck, John Lossing, Owen L. Dawson and Yuan-li Wu (eds.). New York, Praeger, for the Hoover Institution, 1966.
'The core of the technical problems.'

754 KOJIMA, Reiitsu
Recovery of nature: construction in mountain areas before the Great Leap Forward.*Ajia Keizai*, 13(1), 13-35, 1972.
I--The phenomenon of natural disasters. II--New destruction in mountain areas after establishment of new China. III--The initial natural construction plan of the Central Committee. IV--Subsequent movement for construction in mountainous regions. V--Mountain construction and the model of the new mountain economy.

755 *LA NOUVELLE CHINE*
La lutte contre la pollution (The struggle against pollution), No. 8, P. 29-30, Jun 1972.
Multi-purpose use of by-products; removal of factories from urban centers; mass mobilization. The Chinese method is typical of socialists: planning, coordination of enterprises, voluntary labor, mobilization of youth and workers.

756 ORLEANS, Leo A. and SUTTMEIER, Richard P.
The Mao ethic and environmental quality. *Science*, 170(3963), 1173-1176, 1970.
'Maoism is an ethic of progress . . . that appears to make technological development dependent on social development, instead of letting social development slip completely out of phase with technological progress.' The attitudes of the Communist regime toward the natural environment are exploitative--through the fostering of modern science and technology--and curative--through afforestation, water conservancy, land reclamation, and sanitation and public health. Human pollution. Industrial pollution. Two reasons for Chinese Communists' concern with environmental pollution: as part of China's efforts to improve health and sanitation conditions; as part of programs promoting frugality and economy.

757 SIGURDSON, Jon
China: re-cycling that pays. *Läkartidningen*, 69(23), 2837-2841, 1972.
There is no doubt that there is much concern with environmental problems in China; however, priorities and solutions may be different from those of other countries. 'This article is not an analysis of environmental policies . . . but rather an attempt to describe some activities which are likely to have important environmental consequences in China.' Waste disposal; recycling and multi-purpose use; dispersion of industry and local industrial development.

758 SNYDER, Charles
Tomorrow's challenge. *Far Eastern Economic Review*, 70(44), 43-4, 1970.
Continuing lack of economic resources fosters efforts by industry to re-utilize waste products. 'The anti-waste campaign's success indicates the potential for future pollution control in China through mobilization of the masses.' 'Industrialization and the rationalization of farming will without systematic control produce dirtier water and air and poisoned soil.'

See also the following entries:

Technology and Economic Growth. 3.
Technology Policy:
  Choice of Techniques. 40.
  Innovation and the Direction of Technological Change. 67, 74.
Policy for Science. 162, 171.
Policy Toward Science. 218.
Scientific Communication. 377
Agricultural Technology. 465, 469, 470, 471, 472, 483, 485, 495.
Industrial Technology:
  Chemical Industry. 523.

# K Technical Assistance from China

759 GARRATT, Colin
China as a foreign aid donor. *Far Eastern Economic Review*, 21(3), 81, 84-7, 1961.
Technical aid from China.

760 JOHNSTON, Douglas M. and HUNGDAH. Chiu
*Agreements of the People's Republic of China, 1949-1967: a calendar*
Cambridge, Harvard University Press, 1968.
Agreements on aid, economic and technical; scientific and technical cooperation.

761 KOVNER, Milton
Communist China's foreign aid to less developed countries. *An economic profile of Mainland China*, U.S. Congress Joint Economic Committee. New York, Praeger, P. 609-20, 1968.
Technical assistance, p. 614-15.

762 *NOTES ET ETUDES DOCUMENTAIRES*
L'aide bilaterale de la Republique Populaire de Chine aux pays en voie de developpement, No. 3165, 24 Feb 1965. Paris, Secretariat General du Gouvernement, Direction de la Documentation.
Bilateral aid from the PRC to underdeveloped countries--by country.

763 RICHER, Philippe
*La Chine et le Tiers Monde (1949-1969)*. Paris, Payot, 1971.
Pt. 4--Economic and technical aid from the PRC to countries on the way to development, p. 325-46. I--The beneficiary countries. II--Characteristics of the aid; conditions of utilization; psychological aspects; effectiveness and difficulties of Chinese aid. III--Aid: a weapon against the West and the USSR. IV--Model of development?; the Chinese model; reception accorded the aid.

764 TANSKY, Leo
China's foreign aid: the record. *People's Republic of China: an economic assessment*, U.S. Congress: Joint Economic Committee. Washington, U.S. Govt. Printing Office, P. 371-82, 1972. Also *Current Scene*, 10(9), 1-11, 1972.
Including technical assistance.

765 WOLFSTONE, Daniel
Burma's honeymoon with China. *Far Eastern Economic Review*, 33(8), 353-55, 1961.
Technological aid.

See also the following entries:

Technology and Economic Growth. 1, 4, 12, 13, 23.
Technology Policy:
Choice of Techniques. 36, 40.
Innovation and the Direction of Technological Change. 69, 84, 99, 108, 113.

Policy for Science. 118, 119, 123, 129, 130, 135, 141, 146, 154, 155, 206.
Research and Development:
  Military/Space Exploration and Travel/Nuclear Energy. 310, 311, 313, 317, 325, 326, 327, 329, 330, 343, 344, 351, 359, 362.
Higher Education and Technical Training. 436.
Industrial Technology:
  Electric Power Industry. 497, 498.
  Electronics Industry. 505, 512.
  Chemical Industry. 532.
  Petroleum Industry. 533, 538, 539, 540, 546, 551.
  Machinery Industry. 588, 592.
  Basic Metals Industries. 601.
  Mining Industry. 629.
Engineering. 637.

# Appendices

The items included in Appendix I deal with the introduction of post-Renaissance Science and Western technology to China, and their history in China prior to the People's Republic. Appendix II contains items on traditional Chinese science and technology. For other references see John Lust *Index Sinicus* (entry no. 798) and the annual issues of *ISIS: Critical Bibliography of the History of Science and its Cultural Influences* (entry no. 862).

## Appendix I. *Modern Science and Technology in China before 1949.* Entries 766 to 833.

## Appendix II. *Traditional Chinese Science and Technology.* Entries 834 to 944.

766 ADOLPH, William H.
Chemistry in China. *Chemical and Engineering News,* 24(18), 2494-2498, 1946.
Early Chinese chemical industries; silk, paper, printing, porcelain, gunpowder, white lead, zinc, alloys. Alchemy and iatrochemistry in China antedate European developments. The medieval period in Chinese chemistry lasted until the end of the 19th century: in accuracy rather than scientific thinking, discovery by blind experiment. Introduction of Western science and the scientific method de novo in 1910; the author's experiences as a young chemistry instructor in China. Professional associations; journals; research laboratories. Chemistry during World War II.

767 BASALLA, George
The spread of Western science. *Science,* 156(3775), 611-22, 1967. Theory of the development of science. Chinese science and history of Western science in China *inter alia.*

768 BENNETT, Adrian Arthur
John Fryer: The introduction of Western science and technology into nineteenth century China. *Harvard East Asian Monograph Series* No. 24, 1967.
John Fryer . . . 'was for twenty-eight years a translator for a Chinese government arsenal in Shanghai, contributing in his lifetime a total of one hundred and forty-three publications in Chinese, mainly translations of English language works on technology and science.' This volume is a 'reconstruction of John Fryer's life and thought while he was in China and an analysis of the scope of his translated works. I hope also to point out the influence of Fryer's publications on the literati of late nineteenth century China.

769 BERGEMAN, Thomas H.
*The origins of modern science in China: research in chemistry and mathematics before 1938,* unpublished seminar paper. Harvard University n.d.
Before 1929, technology and applied sciences were carried on, while scientifically trained people devoted their efforts to education and government. By the mid-1930s the Chinese scientific community generated certain problems from within from experimental data and theoretical calculations. Only in organic chemistry did Chinese science develop to the point of individual scientific investigation on the basis of an entirely Chinese education before 1938.

770 BERNARD, Henri
*Matteo Ricci's scientific contribution to China.* Peiping, Henri Vetch, 1935.
Chap. I--The legacy of Islam in China and in Europe toward the end of the XV century. II--Ricci's scientific training; Ricci and Chinese science; the problem of Chinese astronomy; the solar eclipse of December 15, 1610. 'Ricci well deserves to be considered as the scientific initiator of modern China.'

771 BIGGERSTAFF, Knight
*The earliest modern government schools in China.* Ithaca, Cornell University Press, 1961.
Introduction of the teaching of science in pre-Republican China.

772 BIGGERSTAFF, Knight
Shanghai Polytechnic Institution and reading room: an attempt to introduce Western science and technology to the Chinese. *Pacific Historical Review,* Vol. 25, P. 127-49, May 1956.
Account of the organization founded by 'well disposed foreigners and progressive Chinese' in 1874 to 'bring Western scientific and technological knowledge to China.'

773 BOSWELL, P.G.H.
Geology in China. *Nature,* 150(3814), 649-50, 1942.
Review of an address by Dr. J.S. Lee, chairman of the Geological Society of China, on the occasion of the Society's 20th anniversary.

774 BRIGHTMAN, R.
Science in China. *Nature,* 163(4149), 704, 1949.
Review of *Science Outpost,* by Joseph Needham and Dorothy Needham (entry no. 819).

775 CH'EN Ch'i-t'ien (CH'EN, Gideon)
*Pioneer promoters of modern industrial technique in China.* Peking, Yenching University, 1934-1938. New York, Paragon Book Reprint Corporation, 1968.
'The purpose of the study is to go over the scattered materials--official documents, correspondence, biographies, and other contemporary writings--in both Chinese and foreign sources with the aim of discovering the attitudes of the responsible statesmen of the day toward the impact of Western machine-civilization, and to determine the policy these men attempted to carry out for China, as well as the results thereof.' 1--Lin Tse-Hsu: pioneer promoter of the adaptation of Western means of maritime defense in China. 2--Tseng Kuo-Fan: pioneer promoter of the steamship in China. 3--Tso Tsung T'ang: pioneer promoter of the modern dockyard and the woollen mill in China.

776 CHESNEAUX, J.
Science in the Far East. *A general history of the sciences; science in the nineteenth century,* Taton, Rene (ed.), Chap. 6 of Pt. VI. London, Readers Union, Thames and Hudson, 1966.
P. 585-9: New scientific trends in China; the missionary influence; official Chinese encouragement of science.

777 CHESNEAUX, J. and NEEDHAM, J.
Science in the Far East from the 16th to the 18th century. *A general history of the sciences; the beginnings of modern science,* Taton, Rene (ed.). London, Readers Union, Thames and Hudson, 1965.
China, p. 586-94: The Jesuit contribution in the 17th and 18th centuries; the spread of Jesuit scientific influence; renaissance of traditional science; internal obstacles to scientific progress.

778 CHIANG Kai-shek
The way and spirit of science. *Nature,* 152(3850), 180-2, 1943.
Abridgement of an address given in 1942, posted in all Chinese Government laboratories and workshops. The scientific approach. The scientific spirit. The scientific system: limitations; management; recording and preparation; division of labour and cooperation; research; experimentation; analysis and statistics; improvements and inventions. Conclusions: 'To be able to shoulder the heavy responsibility of reviving our nation and completing our revolution, we must have at all costs a clear idea of the content and meaning of science; we must propagate the spirit of science; and we must utilize the methods of science; so that one man will be as efficient as ten, and in one day ten days work will be done.'

779 DAGENAIS, F.
*Science in early Republican China: the development of scientific societies, 1914-1927,* unpublished dissertation. Berkeley, University of California, 1964.
The development of science as an institution under the Republic must be delineated and assessed so that a perspective may be had of the Communists' inheritance in 1949. The Science Society of China (1914); The Geological Society of China (1922); other scientific societies 1916-1925; societies founded by foreigners in China. These societies . . . constituted an important social framework in which the scientific brotherhood could function . . . .

780 D'ELIA, Pasquale M.
*Galileo in China.* Cambridge, Harvard University Press, 1960.
Account of the Jesuits' use of Galileo's theories to revise the Chinese calendars.

781 DOUGLAS, Robert K.
The progress of science in China. *Popular Science Review,* Vol. 12, P. 375-85, 1873.
Commentary on the history of Chinese science: the science of numbers; medicine; little known of physiology; geography. 'Within a narrow circle of scholars the sciences will doubtless be more and more cultivated . . . but at present they do not find favour with the governing classes, who in all they do look for some immediate advantage . . . . For science, as science, they have no love. They are willing to use it to serve the ends they wish to gain at the moment, but they are equally willing to discard it as soon as those ends are accomplished.'

782 FAIRBANK, John K.
The influence of modern Western science and technology on Japan and China. *Explorations in entrepreneurial history,* 7(4), 189-204, 1955.
Every aspect of modern history has its technological component or sub-aspect. The impact on China and Japan of the science and technology which came from Western Europe and North America in the latter half of the 19th and early 20th centuries. 1--The sequence of phases in the early borrowing of Western technology. 2--Contrasting speeds of borrowing and institutionalization of

science in China and Japan. 3--The interplay of science and technology with other factors. 4--Key factors affecting technological change. Summary of methodological problems.

783 FAIRBANK, John K. ECKSTEIN, Alexander, YANG, L.S.
Economic change in Early Modern China: an analytic framework. *Economic development and cultural change*, 9(1), 1-26, 1960.
'This paper tries to characterize broadly the process of economic change in China during the century of disturbance which ended with the collapse of the Ch'ing dynasty in 1911. In approaching this task we focus particularly upon the factors that retarded growth. In order to gain perspective upon this century of economic transformation in China and place it in the context of world economic development, we first outline briefly and schematically several paths which industrialization has followed since 1750 in different parts of the world.' Conclusion: the treaty ports are principal centers of change; certain Chinese institutions as factors of retardation.

784 FURTH, Charlotte
Ting Wen-chiang: Science and China's new culture. *Harvard East Asian Series 42*. Cambridge, Harvard University Press, 1970.
Biography of Chinese geologist, 1887-1936. 'In sum, two endeavors dominated Ting's career and gave it unity: first, the struggle to understand modern science and its impact upon the old Chinese intellectual order; and second, the search for methods of meaningful action on the part of an elite still haunted by the imperatives of Confucian scholar statesmanship.' 'The first four chapters of the book deal largely with Ting's purely scientific career; his political beliefs and actions are the subject of the last three.' Bibliography.

785 GREGORY, J.W.
The scientific renaissance in China. *Nature*, 113(2827), 17-19, 1924.
Description of universities and scientific institutions and societies in China in the 1920s: Union Medical College in Pekin, Hong Kong University, National University of Pekin, Tsing Hua College, Shansi University, Geological Survey and Museum, Archaeological Society, Geological Society of China. 'These noble schemes have been to a large extent frustrated by the political disorders that followed the revolution.'

786 HOPKINS, C.E. and STEPANEK, J.E.
China's AIS--A point four pioneer; tested method of increasing output in underdeveloped lands by putting better tools in hands of average peasant. *Far Eastern Survey*, 18(14), 157-61, 1949.
Description of the Agricultural Industry Service program in China, ca. 1947-48, 'an UNRRA legacy to the government of nationalist China.' 'Its basic idea was to assist and expedite the transformation from primitive agricultural to a modern mixed agricultural-industrial economy . . . by establishing key industrial plants in rural areas.' AIS sought 'to apply the best modern scientific methods toward basic self-perpetuating solutions to China's economic problems.' Organizational pattern

included establishment of central development stations in each of five economic regions, at which industrial and agricultural research and development were performed; extension workers would disseminate improvements and feed in requests for assistance on specific problems. Difficulties encountered . . . include the social and political position of Chinese technicians.

787 HOU, Chi-ming
Economic dualism: the case of China, 1840-1937. *Journal of Economic History*, 23(3), 277-97, 1963.
There is little factual support for the common belief that the traditional sector of the Chinese economy suffered a general decline because of the development of the modern sector. The traditional sector was still predominant in the 1930s. The prolonged coexistence between the traditional and the modern sector may be explained, in large part, by factor prices and factor proportions dictated by technology.

788 HOU, Chi-ming
External trade, foreign investment, and domestic development: the Chinese experience, 1840-1937. *Economic development and cultural change*, 10(1), 21-41, 1961.
'It is our main contention that, in the case of China, certain beneficial links may be found between external economic contact and domestic development . . . . Development in China was basically a matter of change in the . . . attitude of the government toward economic affairs, the direction of employing savings, the level of technology, inventiveness, etc. . . . . If external trade and foreign investment contributed to these preconditions . . . in the Chinese economy, it seems arbitrary to assert that it is the capital-exporting countries that obtained most of the benefits from the economic intercourse . . . .' This is in contradiction to the 'absorption thesis', which states that native entrepreneurial initiative and domestic investment were absorbed in the export sector, and that benefits of export development are absorbed by the investing countries. Historical survey of the record of external trade; the amount of foreign investment; foreign borrowing by the Chinese government; foreign direct investment, and the terms of trade. Indicators of economic modernization show a 'significant trend toward "economic modernization"': the 'retaliation effect'; the 'training ground and imitation effect'; the 'linkage effects'; of foreign trade on Chinese economic modernization, and the 'reinvestment ratio'. Available evidence does not support the argument that Chinese firms suffered 'oppression' as a result of foreign investment.'

789 HOU, Chi-ming
*Foreign investment and economic development in China, 1840-1937.* Cambridge, Harvard University Press, 1965.
'The most obvious link between foreign investment and China's economic modernization was that the former not only performed the pioneering entrepreneurial function of introducing modern technology into a number of fields but also accounted for a large share of the modern sector of the economy.' 1--A general picture of foreign investment in China. 2--Foreign obligations

of the Chinese government. 3--Foreign direct investments in China. 4--Framework of analysis of the effects of foreign investment. 5--Determinants of foreign investment in China. 6--Foreign investment and economic modernization. 7--A dualistic economy. 8--External aspects of foreign investment in China. 9--Summary and conclusion.

790 HSU, Francis L.K.
A cholera epidemic in a Chinese town. *Health, culture, and community: case studies of public reaction to health programs*, Paul, Benjamin D. (ed.). New York, Russell Sage Foundation, P. 135-49, 1955.
'. . . the distinction made by Westerners between "magical" and "scientific" practices is not relevant to rural Chinese . . . . Any attempt to introduce new knowledge or new techniques in a foreign setting will benefit from the realization that all communities respond to these attempts according to premises implicit in their own cultural traditions.'

791 HSU, Francis L.K.
*Religion, science, and human crises: a study of China in transition, and its implications for the West*. London, Routledge and Kegan Paul, 1952.
In any culture, magic and real knowledge are intertwined because man's behaviour is dictated by faith developed out of the pattern of his culture. Case study of a cholera epidemic in a Chinese town.

792 HUGHES, E.R.
*The invasion of China by the Western World*. London, Adam and Charles Black, 1937.
Chap. V--Western science and medicine.

793 KUO, Zing Yang
Reconstruction in China. *Nature*, 149(3767), 42-3, 1942.
'At present there is already great shortage of trained personnel in various fields, especially in science and technology . . . . The training of scientific and technical personnel for cultural and economic reconstruction in China will depend, more than ever before, upon sending students to study in American and British universities.' Response of British and American institutions to the proposals of the Director of the China Institute of Physiology and Psychology.

794 KWEI, Chi-Ting
The status of physics in China. *The American Journal of Physics*, 12(1), 13-18, 1944.
Training of Chinese in physics in America and Britain. Development of physics programs at Southeastern University, Tsing Hua and Peita. Physics curricula. Effect of the war on graduate training. Research institutes and their organization; research programs. Professional associations and publications.

795 KWOK, D.W.Y.
*Scientism in Chinese thought 1900-1950*. New Haven, Yale University Press, 1965.
'Proponents of the scientific outlook in China were not always scientists . . . they were intellectuals interested in using science, and the values and assumptions to which it had given rise, to discredit and eventually to replace the traditional body of values. Scientism can thus be considered as the tendency to use the respectability of science in areas having little bearing on science itself . . . .' 'Many leading thinkers of modern China failed to distinguish between the critical attitude and methodological authority, between scientific objectivity and absolute rationality, and between scientific laws and irrefutable dogmas. This failure left behind an era of open contests of ideas. It helped initiate the next era, a monolithic intellectual supersystem.'

796 LATTIMORE, Owen
The industrial impact on China, 1800-1950. *First International Conference of Economic History: Contributions*. Stockholm, P. 103-13, 1960.
'The reasons for China's failure to go on from high pre-industrial achievement to a process of industrialization . . . (were perhaps the lack of) a suitable conjunction of social structure and profit incentives, or opportunities, favoring the emergence of a class interested in promoting production by labor-saving devices.' Traditional economic structure; the arrival of Western traders in China. The Chinese attempt to adapt foreign technology without accepting foreign ways of thinking.

797 *THE LISTENER*
Science in China, Vol. 33, P. 9-10, Jan 1945.
Quoting Dr. Joseph Needham on his observation of research programs in scientific institutions in wartime China.

798 LUST, John
*Index Sinicus*, a catalogue of articles relating to China in periodicals and other collective publications, 1920-1955. Cambridge, W. Heffer & Sons, Ltd., 1964.
Chap. XVIII--Science and technology. General: I--General; II--History. Mathematics and metrology. Astronomy and calendar. Chemistry. Geology. Botany. Biology. Technology: I--History; II--Textiles; III--Agricultural techniques and history; IV--Shipbuilding; V--Modern industry.

799 MURPHEY, Rhoads
The treaty ports and China's modernization: what went wrong? *Michigan Paper in Chinese Studies No. 7*. Ann Arbor, The University of Michigan, Center for Chinese Studies, 1970.
The role of the treaty ports in the modernization and Westernization of China. 'The port cities which grew out of a few of the earlier European trade bases constituted a working model of dynamic nineteenth-century Europe . . . and functioned as points of entry into Asia of an alien Western order at a time when the

West was in the midst of vigorous economic (technological) institutional growth and when . . . most of Asia was experiencing a period of relative deterioration in all of these respects.' Section V: Trade and technology: Coals to Newcastle? The Chinese resisted Western technology because it was not seen as more advantageous than the traditional pragmatic system, not simply because it was foreign. Foreign technology failed to transform China, but was ideologically important in its revelation of China's backwardness.

800 *NATURE*
Biological research in China, 149(3778), 353, 1942.
Report on research by the National Institute of Zoology and Botany of the Academia Sinica--as recorded in *Sinensia*, Sep 1940.

801 . *NATURE*
Chemistry in China, 156(3965), 511, 1945.
Note on publication of *Journal of the Chinese Chemical Society*, 1941-1944. 'A large number of papers are of general chemical interest, as distinguished from those concerned with national requirements, and a very good balance is preserved between the various branches of investigation . . . . The standard of the papers is high . . . .'

802 *NATURE*
Geographical research in China, 154(3910), 458, 1944.
Review of an article by Professor Chi-Yun Chang in *Annals of the Association of American Geographers*, 39(1), 1949.

803 *NATURE*
Mathematics in China, 154(3920), 763, 1944.
Announcement of the work of Professor L.K. Hua. 'In addition to work directly for the war effort and in spite of difficulties of communication, mathematicians in China are able to produce a considerable amount of new work of the highest quality.'

804 *NATURE*
Medical progress in China, 149(3784), 523, 1942.
'In a recent lecture published in the *Asiatic Review* of April, Dr. W.H. Woo gives an interesting survey of medical progress in China from the earliest times.'

805 *NATURE*
Message from Chinese men of science, 151(3839), 612, 1943.
Open letter to British scientific men from Professor Tseng Chao-lun, head of Department of Chemistry, National Southwest Associated University, Kunming, in response to visit by British Council cultural mission.

806 *NATURE*
Parasitology in Free China, 152(3867), 699-700, 1943.
Report of the contents of an article by K. Chang in a typewritten copy of *Acta Brevia Sinensia* No. 4, transmitted by the British Council. 'It is evident that parasitologists of Free China are doing admirable work under very difficult conditions.'

807 *NATURE*
Science and engineering in China, 149(3790), 693, 1942.
Review of a pamphlet, *China today: the thirtieth anniversary of the Chinese Republic, 1911-1941*, published by Chinese students in Britain (1942?). Contains three articles: by P.M. Yap, describing utilization of science in China; an article which indicates that traditional medicine is detrimental to development of the medical profession; and one by T.C. Chan, reviewing Chinese engineering.

808 *NATURE*
Scientific development in China, 160(4070), 600, 1947.
Resolutions passed by the Chinese Association for the Advancement of Science and other scientific societies, 30 Aug/1 Sep 1947 on atomic research and scientific research in China. Scientific backwardness is a cause of poverty in China; sufficient emphasis has not been given to fundamental sciences; China should adopt a long-range plan for scientific development.

809 *NATURE*
Seismology in China, 154(3907), 360, 1944.
Active recording of earthquakes continues despite war-time disruption.

810 NEEDHAM, Joseph
Chinese astronomy and the Jesuit Mission. *China Society Occasional Papers No. 10*. London, The China Society, 1958.
Paper read to the China Society, 20 Mar 1957, based on *Science and civilization in China*, Vol. 3. Transmission to China by Jesuit missionaries in the +17th century of Western astronomy.

811 NEEDHAM, Joseph
*Chinese Science*. London, Pilot Press Ltd., 1945.
Photographs from the author's post in China at the British Council Cultural Scientific Office during World War II. Preface discusses scientific research during the war, organization for teaching and research, research projects, medical science.

812 NEEDHAM, Joseph
Chungking Industrial and Mining Exhibition. *Nature*, 153(3878), 247, 1944.
Exhibition of technology organized by National Resources Commission in March 1944. 'Signalizes in a striking way the determination of China to embark upon large-scale industrialization.'

813 NEEDHAM, Joseph
Science and technology in China's far south-east. *Nature*, 157(3981), 175-8, 1946.
Scientific institutions in Kuangtung and Fukien, as observed by the author, a member of the British Scientific Mission in China.

814 NEEDHAM, Joseph
Science and technology in the north-west of China. *Nature*, 153(3878), 238-41, 1944.
Conditions of scientific work in Shensi and Kansu during World War II, as observed by the author while a member of the British Council Cultural Scientific Mission in China.

815 NEEDHAM, Joseph
Science in Chungking. *Nature*, 152(3846), 64-66, 1943.
Scientific institutions in the Chungking region during World War II.

816 NEEDHAM, Joseph
Science in Kweichow and Kuangsi. *Nature*, 156(3965), 496-99, 1945.
Scientific conditions in China's far south-east as observed by the author as a member of the British Scientific Mission in China.

817 NEEDHAM, Joseph
Science in South-West China. *Nature*, 152(3844), 9-10, 1943, I--The physico-chemical sciences; 152(3845), 36-7, 1943, II--The biological and social sciences.
The author, a member of the British Council Scientific Mission in China, describes conditions of scientific institutions in wartime Kunming.

818 NEEDHAM, Joseph
Science in Western Szechuan. *Nature*, 152(3856), 343-5, 1943, Pt. I--Physico-chemical sciences and technology; 152(3857), 372-4, 1943, Pt. II--Biological and social science.
Scientific institutions in wartime Szechuan, described by the author, a member of the British Council Scientific Mission in China.

819 NEEDHAM, Joseph and Dorothy
*Science Outpost*. Papers of the Sino-British Science Cooperation Office (British Council Scientific Office in China) 1942-1946. London, Pilot Press Ltd., 1948.
'The Sino-British Science Cooperation Office . . . was part of the Allied attempt to break the Japanese intellectual and technical blockade round China. It was our aim to bring help to the Chinese scientists and technologists isolated in the cities . . . and the remoter locations where war factories had grown up and exiled universities had established themselves.'

820 PEAKE, Cyrus H.
Some aspects of the introduction of modern science into China. *ISIS*, Vol. 22, P. 173-219, 1934.
'It is the purpose of this paper to outline the history of modern science in China during the three and a half centuries since it was first introduced by the Jesuits in the seventeenth century' . . . in four historical periods, 1600-1800; 1800-1895; 1898-1912; 1911-1930.

821 *SCIENCE AND TECHNOLOGY IN CHINA*
National Academy of Peiping, 1(6), 101-5, 1948.
Organization and personnel of the National Academy of Peiping; research-in-progress in the Institute of Physics, Institute of Atomic Physics; Institute of Chemistry, Institute of Materia Medica, Institute of Physiology, Institute of Zoology, Institute of Botany and Institute of Historical Studies and Archaeology.

822 SEWELL, W.G.
A chemist in China. *Chemistry in Britain,* 8(12), 529-33, 1972.
Memoirs of a British chemist who taught at Chengtu University from 1924 to 1952, with comments on science in China today. 'We may find it difficult to understand how chemical education and research now function, but the challenge of their uncompetitive system is something to be watched.'

823 SHEEKS, Robert
*Science in China today.*
Source ?
History of Chinese science and technology briefly reviewed; fundamental scientific discoveries and technological inventions had little effect on traditional China. Matteo Ricci, 17th century; modern science for defense introduced in 19th century; influence of Western-educated Chinese in the Republic after 1911; after 1949, the classic pattern of the previous three waves persisted; the Chinese Communists took great interest in military technology, accepted science as an adjunct to that technology, but continued to resist ideological, foreign, and intellectual influences.' Review of science in Communist China. The fifth wave of science now rising from within as Chinese scientists look upon science as their own.

824 SILOW, R.A.
The scientific activities of the British Council in China.
*Science and technology in China,* 1(2), 26-32, 1948.
Account of the history and activities of the British Council in China during and after World War II. The aims of the British Council.

825 SIVIN, Nathan
On 'China's opposition to western science during late Ming and early Ch'ing'. *ISIS,* Vol. 56, P. 201-5, 1965.
The author contributes corrections to 'several errors of translation, transliteration, documentation and fact' in George H.C. Wong's 'China's opposition to western science during late Ming and early Ch'ing'. *ISIS,* Vol. 54, P. 29-49, 1963 (entry no. 832).

826 SPENCE, Jonathan
*The China helpers: Western advisers in China, 1620-1960.* London, Bodley Head, 1969.
Biographies of sixteen Western advisers in China between the 1620s and 1950s including engineers, doctors.

827 TANG, Pei-sung
Biology in war-time China. *Nature*, 154(3897), 43-6, 1944.
'This article will be of the nature of a report on the movements of biological institutions during the war and the activities of biologists associated with those institutions.'

828 TAWNEY, R.H.
*Land and labour in China*. London, George Allen & Unwin Ltd., 1932.
'The influence of Western science, technology, economic conceptions, social theories and ethical standards on a civilization which, half a century ago, if aware of them at all, regarded them with repugnance, is obviously a phenomenon of profound significance.' China's historical contributions to science and technology; the present contrast between China and the West in 'technological equipment and industrial organization.' Chap IV--The possibilties of rural progress: Section IV--Science and education; VIII--Population, migration and the development of industry. Chap. V--The old industrial order and the new.

829 TING, V.K.
Scientific research in China. *Nature*, 136(3432), 208-11, 1935.
The Secretary-General describes the organization and the Academia Sinica; its relationship to the National Government of China; the structure and scientific activities of its constituent institutes; its relations to international science; its budget.

830 WALLACE, Henry A.
The U.S., the U.N. and Far Eastern agriculture. *Bulletin of the Atomic Scientists*, 6(12), 363-4, 1950.
'(Chinese farmers) have reached the conclusion that they must utilize modern technology both in agriculture and in industry in order to survive. The question is: will it be Russian or Anglo-American technology or both? . . . . I am confident that the Russians cannot do as good a job in helping the Chinese with their agriculture as we.' Main technological aspects of agrarian reform: 1--use of phosphate fertilizers; 2--adaptation of machinery for small farms; 3--flood control and irrigation; 4--erosion control; 5--improvement of plant and livestock strains; 6--improvement of marketing; 7--control of plant and animal disease.

831 WINFIELD, Gerald F.
*China: the land and the people*. New York, Sloane, 1948.
China's public health problem. Agricultural techniques. Population.

832 WONG, George H.C.
China's opposition to Western science during Late Ming and Early Ch'ing. *ISIS*, Vol. 54, P. 29-49, 1963.
Examination of Chinese scholar-officials' opposition to Western science primarily in terms of basic traditional beliefs and ancient authorities. Owing to the slow development of native science and the low state into which it had fallen during the Ming dynasty, there appeared to be a partial Chinese acceptance of

Western science and techniques--and a semi-popular movement to re-examine and re-evaluate Chinese science. Skeptical, xenophobic anti-Western scholars attempted to isolate traditional Chinese science from Western scientific influence and to negate the Jesuits' use of their scientific knowledge as a means of reaching their religious objectives. Involved in this movement was a denial of the superiority of various parts of Western science--with the contention that all Western scientific ideas and techniques had originated in China. See also entry for Sivin, Nathan, *ISIS*, Vol. 56, P. 201-5, 1965 (entry no. 825).

833 YAP, Pow-meng
*The place of science in China*. London, The China Campaign Committee, n.d. (1944?*).
Chap. I--Science in Ancient China: 'If the practice of experimentation--the essential part of the scientific method as we know it--failed to establish itself in the intellectual tradition of China, it was largely because of social and economic reasons.' Chap. II--Science in Modern China: scientific research institutions; the Academia Sinica; the National Academy of Peiping; the Science Society of China; the Natural Science Society of China; the Fan Memorial Institute of Biology; Henry Lester Institute for Medical Research; technical research institutions founded by private interests. Chap. III--Chinese Science and the war: the National Economic Council; the Ministry of Economic Affairs; the National Resources Commission; the Ministry of Agriculture and Forestry; the National Geological Survey; the National Health Administration; science education in wartime. Chap. IV--Science and the ideological struggle in the East: 'The legacy of objectiveness and rationalism . . . has saved modern China from much of the frustration of contemporary Japan.' *See review in *Nature*, 153(3878), 247, 1944.

834 BEER, A., HO, Ping-yü, LU, Gwei-Djen, NEEDHAM, Joseph, PULLEYBLANK, E.G., THOMPSON, G.I.
An 8th-century Meridian Line: I-Hsing's chain of gnomons and the pre-history of the metric system. *Vistas in astronomy*, Beer, Arthur (ed.). Oxford, Pergamon, Vol. 4, P. 3-29, 1961.
Study of systematic observations and measurement of solstitial and equinoctial sun shadows and of polar altitudes in 725 A.D. by a Buddhist monk (I-Hsing) and an official astronomer. The ratio of terrestrial distance units to the degree thus ascertained fixed a civil unit in a manner prefiguring the metric system. Translation of original texts. I-Hsing's texts do not indicate whether he attempted to derive from his measurements a value for the circumference of a spherical Earth, but there is little reason to suspect Confucian orthodoxy of suppressing such a notion. 'I'Hsing's work takes an outstanding place in the pre-history of the metric system.'

835 BERNAL, J.D.
*Science in History*. London, Watts and Co., 1965.
P. 230-42, Technical innovations from the East and China (in the Middle Ages); P. 883-6, The Chinese Revolution.

836 CHESNEAUX, Jean
Le 'miracle chinois'. *La Recherche*, 3(23), 420, 1972.
History of development of science and technology in China, based on the work by Joseph Needham.

837 COMBRIDGE, J.H.
The Chinese water-balance escapement. *Nature*, 204(4964), 1175-1177, 1964.
History of technology.

838 DAVIS, Tenney L. and CH'EN Kuo-fu
Shang yang-tzu, Taoist writer and commentator on alchemy. *Harvard Journal of Asian Studies*, Vol. 7, P. 126-9, 1942.
Shang yang-tzu (late Sung-early Yuan) was author of a picture and diagram, and Chin tan ta yao, 'which appear to prove . . . that the essential doctrine of Chinese alchemy is the same as that of the alchemy of Europe.'

839 DIETRICH, Craig
Cotton culture and manufacture in Early Ch'ing China. *Economic organization in Chinese society*, Willmott, W.E. (ed.). Stanford, Stanford University Press, P. 109-35, 1972.
Part of the cotton industry in Ch'ing China was differentiated and adaptive, possessing a range of techniques and organizational forms which permitted cloth to be made both by self-sufficient families and by a system of market-oriented specialists. Description of cotton technology as it existed in Ch'ing China. Conclusion: the technology of cotton culture and manufacture was pre-modern.

840 EBERHARD, Wolfram
The political function of astronomy and astronomers in Han China. *Chinese thought and institutions*, John K. Fairbank (ed.). Chicago, The University of Chicago Press, P. 33-70, 1957.
1--The question of despotism and its limitations in Ancient China. 2--The data concerning natural phenomena. 3--The political character of the calendar. 4--Factors militating against a development of science: Chinese astronomers were not interested in pure science nor in applied technical sciences, but in politics. Improvement of calendars was regarded as a revolutionary act by the Han rulers; hence Chinese scientist-astronomers became the nucleus of anti-dynastic and revolutionary movements and were eventually crushed. 5--The domination of political interest.

841 ELVIN, Mark
The high-level equilibrium trap: the causes of the decline of invention in the traditional Chinese textile industries. *Economic organization in Chinese society*, Willmott, W.E. (ed.). Stanford, Stanford University Press, P. 137-72, 1972.
'The basic idea is that stimulus to invention in the economic field usually takes the form of a change in the pattern of supply and demand . . . . In most cases . . . the effect of the response will be to alter yet again the pattern of supply and demand . . . . But if, for whatever reason, the flow of such causally interlocking changes comes to an end, there will also be an end to economically significant invention. It will not reappear until some extraneous event . . . disrupts the equilibrium.' At the point of the high-level equilibrium trap, 'increased inputs of labor, capital, and organization yield no returns. Pre-modern technology and practice are both at a maximum.' Example: the inhibition of mechanical invention in the Chinese cotton industry, due to the high-level equilibrium trap in which the Chinese economy was caught between the 14th and the 18th centuries.

842 ELVIN, Mark
*The State, printing and the spread of scientific and technical knowledge in China, 950-1350*. XIII International Congress of the History of Science, Moscow 18-24 Aug 1971. Moscow, Nauka Publishing House, 1971.
'The purpose of this paper is to put forward some suggestions as to the reasons for the vigorous growth of scientific and technical knowledge in China (from 950 to 1350) and to intensify rather than to solve the problem of why Chinese scientific creativity declined so greatly after about the middle of the fourteenth century.' '. . . two features in particular distinguished medieval China from the medieval West. There were the existence of an imperial state which pursued a consistent policy of spreading and advancing scientific and technical knowledge; and the art of woodblock printing or xylography . . . ..' Conclusion: '. . . even printing and the educational efforts of the state had not in fact removed a strong regional character from much of the Chinese science, and . . . this may have made it vulnerable.' 'It therefore seems possible that the chaos into which much of Northern China fell during the period when Mongol rule there was collapsing may

have irremediably damaged a distinctive scientific tradition with a stronger bent towards the theoretical than its southern counterpart.' '. . . the fourteenth century must have in some way marked a divide in the way in which the Chinese looked at the world . . . a movement away from the conceptual mastery of external nature and a movement towards introspection and subjectivity. The consequences for science were probably major.'

843 ENNIS, Thomas E.
The role of Chinese science and technology in modern civilization. *Eastern World*, 20(5/6), 15-16, 29, 1966; 20(7/8), 15-16, 1966.
Brief survey of technological achievements in traditional China, based on four volumes of *Science and civilization in China*, (Needham). 'The unprogressive character of Chinese science is due to several peculiar conditions . . . . In order to obtain full understanding, the concepts of Yin and Yang and feng shui must be examined.'

844 FUNG, Yu-lan
Why China has no science--an interpretation of the history and consequences of Chinese philosophy. *The International Journal of Ethics*, 32(3), 237-63, 1922.
'China produced her philosophy at the same time with, or a little before, the height of Athenian culture. Why did she not produce science at the same time with, or even before, the beginning of modern Europe? This paper is an attempt to answer this question in terms of China herself.' '. . . I shall venture to draw the conclusion that China has no science, because according to her own standard of value she does not need any.' Review of major schools of Chinese philosophy: 'Taoism stood for nature as against art'; 'Maoism stood for art as over against nature'; 'Confucianism is a mean between the two extreme standpoints of nature and art.' Confucius himself and Mencius stood nearer to the extreme of nature; Suen Tse stood nearer to that of art . . . (but) 'Suen Tse's teaching, together with the Chin dynasty, disappeared soon and forever.' Sung Neo Confucianism: 'This period of the history of Chinese philosophy was almost perfectly analogous to that of the development of modern science in European history, in that its productions became more and more technical, and had an empirical basis and an applied side. The only, but important, difference was that in Europe the technique developed was for knowing and controlling matter, while in China that developed was for knowing and controlling the mind.'

845 GALLAGHER, Louis J.
*China in the 16th century: the journals of Matthew Ricci*. New York, Random House, P. 1583-1610, 1942.
'The author of numerous works on science and religion, written in Chinese, Ricci was well known to the educated classes of China as a prominent professor of physics, mathematics and geography, as a learned philosopher of Chinese and of extraneous doctrine, as a prominent commentator on Confucius, and particularly as an eminent teacher of the Christian religion.' Book One

includes: 4--concerning the mechanical arts among the Chinese; 5--concerning the liberal arts, the sciences, and the use of academic degrees among the Chinese.

846 HAUDRICOURT, A. and NEEDHAM, Joseph
Ancient Chinese science, Chap. 5 of Pt. I, P. 161-77. Science in Medieval China, Chap. 4 of Pt. III, P. 427-439 in *A General History of the Sciences, Ancient and Medieval Science*, Taton, Rene (ed.). London, Readers Union, Thames and Hudson, 1965.
Ancient Chinese Science: historical background; mathematics; astronomy; physics and general biology. Science in Medieval China: mathematics; astronomy and geography; natural science.

847 HO, Peng-yoke
Ancient Chinese astronomical records and their modern applications. *Physics Bulletin*, Vol. 21, P. 260-63, Jun 1970.
The place of astronomy in ancient China; astronomical records and their preservation; Korean, Japanese and Vietnamese astronomical records; East Asian astronomical records in modern research. Applications of ancient and medieval East Asian astronomical records to the study of novae and supernovae; pulsars; comets and meteor showers; spectrum of time.

848 HO, Peng-yoke
The birth of modern science in China, inaugural lecture at the University of Malaya, 25 Nov 1966. Kuala Lumpur, 1967.
'The main purpose of this lecture is to discuss the failure of traditional Chinese science in giving birth to modern science until the time when it got engulfed into the mainstream of our modern universally valid world-science.' Inhibiting factors are classified as the difficulties inherent in the logic or order which formed the basic scientific ideas worked out by the ancient Chinese, political and social factors, and inter-cultural transmission with neighbouring countries. Although there were signs of a Renaissance in Chinese science in the 17th and 18th centuries, after the Opium War Chinese science began to merge into the mainstream of modern science.

849 HO, Ping-yu and NEEDHAM, Joseph
Ancient Chinese observations of solar haloes and parhelia. *Weather*, 14(4), 124-34, 1959.
Observations of solar haloes and parhelia began in China at about the same time as the studies of the Peripatetic school in the West, but were much more thoroughly recorded and by the +7th century constituted a body of knowledge unapproached until the Renaissance in the West a thousand years later.

850 HO, Ping-yu and NEEDHAM, Joseph
Elixir poisoning in Mediaeval China. *Janus*, 48(4), 221, 1959. Also in *Clerks and craftsmen in China and the West*. Cambridge, Cambridge University Press, P. 316-39, 1970.
Elixir poisoning, its repercussions and the countermeasures for its prevention taken by the Chinese alchemists themselves.

851 HO, Ping-yu and NEEDHAM, Joseph
The laboratory equipment of the Early Mediaeval Chinese alchemists. *Ambix*, 7(2), 1959.

852 HO, Ping-yu and NEEDHAM, Joseph
Theories of categories in Early Mediaeval Chinese alchemy. *Journal of the Warburg and Courtauld Institutes*, 22(3/4), 1959.
The contributions of Chinese alchemists to chemical discovery and invention were made in a largely empirical tradition, but they were by no means strangers to theoretical formulations, even though these remained until the end of a typically pre-Renaissance character. Full translation of Tshan Thung Chhi Wu Hsiang Lei Pi Yao, a Thang text with Sung commentary, containing 'a conscious body of theory.' List of categories in early mediaeval Chinese alchemy (in Chinese).

853 HODGES, Henry.
*Technology in the Ancient World*. London, Allen Lane, The Penguin Press, 1970.
'This book is intended to be a straightforward account of the development of mankind's technology . . . written specifically with the general reader in mind . . .' P. 217-30: 'The barbarians in the East: China.' Technologies for potting, bronze-founding, the chariot and the composite bow were borrowed from Western Asiatic societies. Porcelain, writing. Iron techniques, the cross-bow, true draught harness. 'Although there seem to be many points of similarity between the development of technology in China and in Western Asia during this period, there was a very marked difference in the organization of workshops and industry generally.'

854 HSIEH, Chiao-min
Hsia-ke Hsu--Pioneer of modern geography in China. *Annals of the Association of American Geographers*, 48(1), 73-82, 1958.
Biography of Hsia-ke Hsu (1586-1641), 'the first Chinese to make long treks in the interest of scientific discovery and geographical investigation.' Evaluation of Hsu's contribution to the development of the science of geography in China: 'Hsu's startlingly modern concept of securing geographical data from the field was an important step in the development of geography in China. Keen observation and detailed recording . . . were fully recognized and employed by Hsu.'

855 HU, Shih
*The Chinese Renaissance*. New York, Paragon Book Reprint Corp., 1963 (first published University of Chicago, 1934).
Originally delivered as the Haskell Lectures of the Department of Comparative Religion, University of Chicago, July 1933. I--Types of cultural response: comparison of Japanese and Chinese response to initial contact with the West, including Western technology. II--Resistance, enthusiastic appreciation, and the new doubt: changes in Chinese conceptions of Western civilization: Chinese first impressed by the Jesuits' scientific accomplishments. After defeats beginning with the Opium Wars, self-reproach and appreciation of Western civilization--including

science and technology, although this coincided with the 'transvaluation of values in the Western civilization' after the first World War. IV--Intellectual life, past and present: '. . . the intellectual life in China has been confined to the sphere of ethical, social, and political philosophy, and the purely literary training of the intellectual class has tended to limit its activities to the field of books and documents . . . in the last 800 years, there has grown up a scientific tradition, first as an intellectual ideal taught in the most influential school of philosophical thought, then as a scientific technique, even though it was applied not to the objects of natures but to humanistic and historical studies.' Summary history of the schools of Chinese philosophy.

856 HU, Shih
The scientific spirit and method in Chinese philosophy. *The Chinese Mind,* Moore, Charles A. (ed.). Honolulu, East-West Center Press, P. 104-31, 1967.
An 'historical approach to the comparative study of philosophy': intellectual activities are shaped by geographical, climatic, economic, social, political and individual biographical factors, all capable of being understood; in the historical development of science, the scientific spirit or attitude of mind and the scientific method are of more importance than practical or empirical results. The scientific spirit and method in the intellectual and philosophical history of China, defined as 'dispassionate search for the truth'; 'critical use of human reason'; 'disciplined intellectual inquiry'; 'exact and impartial inquiry'.

857 HUARD, P. and WONG, M.
Les enquetes francaises sur la science et la technologie chinoises au xviiie siecle. *Ecole francaise d'Extreme-Orient, Bulletin,* Vol. 53, P. 137-226, 1966.
Historical survey of Western views on Chinese science and technology. I--The historical context: the 18th century was a century of sinophilia, sinomania, and sinophobia, and also, for the first time, the Chinese race and certain important elements of the Chinese culture were recognized. II--Views of the French Government. III--Collectors of documents: the missionaries; the French at Canton. IV--Scientific surveys (based on the memoires of 27 missionaries): astronomy and mathematics; physics and meteorology; geography; ethnography; zoology; botany; medicine. V--Survey of technology: agriculture; horticulture; urban and rural architecture; military art; ceramic and other arts; chemical industry; lacquer; metallurgy; music; paper and printing; weights and measures; textiles; transport. Extensive bibliography and footnotes.

858 HUARD, P. and WONG, M.
Evolution de la Matiere Medicale Chinoise. *Janus,* Vol. 47, 1958.
Chinese medicine has therapeutic qualities very different from those of the Mediterranean regions. We think it best to group the principle works on medical materials in relation to their different epoques. Because of its importance the Pen-ts'ao Kang-mou of Li Che-chen (1518-1593) will be another study in itself.

Our plan will be the following: The principle phases: to the origins in Han (1500 B.C. to 206 A.D.); Han (206-220); Suei and Tang (581-907); Sung (960-1279); Chin-Yuan (1206-1368); Ming (1368-1644); Ching (1644-1912). The connection between Chinese medical materials and other Eurasian medical materials.

859 HUARD, Pierre and WONG, Ming
Le developpement de la technologie dans la Chine du XIXe siecle. *Cahiers d'Histoire Mondiale,* 7(1), 68-85, 1962.
Traditional science and technology: agriculture; animal husbandry; mining; metallurgy; public works; ceramics; postal service. Western technology in China (especially in the 19th century).

860 HUARD, Pierre and WONG, Ming
*Chine, d'hier et d'aujourd'hui.* Paris, Horizons de France, 1960.
Chap. 12--Chinese sciences and technologies: astronomy; chemistry and alchemy; zoology; mathematics; geology; medicine; physics; botany; technology. Brief reviews (1-2 paragraphs) of Chinese sciences and technology. Illustrations: photographs and reproductions of Chinese drawings and diagrams.

861 HUARD, Pierre and WONG, Ming
*Le Taoisme et la science.* Florence, Actes du VIIIeme congress international d'histoire des sciences, P. 1096-1098, 1956.
Taoist thought contributed to chemistry, medicine and mathematics, and was scientific and revolutionary in character, containing, in embryo, natural philosophy and evolution.

862 ISIS
*Critical bibliography of the history of science and its cultural influences* (Annual).
Entry No. HS35.2: The Far East (to ca. 1600): General histories and histories of science; The exact sciences; Natural history; Pseudo-science and experiment; Technology, travel, exploration and geography; Medicine and health.

863 KIANG, T.
Possible dates of birth of pulsars from ancient Chinese records. *Nature,* 223(5206), 599-601, 1969.

864 KU, Y.H.
A survey of Chinese culture--science. *Symposium on Chinese Culture,* Cheng Chi-pao (ed.). New York, China Institute in America and American Association of Teachers of Chinese Language and Culture, P. 28-33, 1964.
Pt. I--China's past scientific developments and their contributions to the world: astronomy; mathematics; measures and weights and their standardization. Pt. II--Chinese intellectuality and scientific development: mathematics; chemistry; metallurgy.
Pt. III--Influence of Chinese cultural background upon modern scientific development.

865 LU, Gwei-Djen
China's greatest naturalist; a brief biography of Li Shih-Chen. *Phycis,* Vol. 8, P. 383-92, 1966.

'Li Shih-Chen . . . 1518 to 1593, the greatest naturalist in Chinese history . . . his scholarly attitude to the wealth of previous literature makes him also the greatest Chinese historian of science before modern times, and his works are an unparalleled source of information on the development of scientific knowledge in East Asia.' Li Shih-Chen's scientific attainments were in developing systematic principles of classification; in the biological field; in the fields of chemistry and mineralogy. His greatness as a naturalist was not recognized during his lifetime, but he was well known as a physician. His (indirect) influence on modern scientific development in Europe.

866 LU, Gwei-djen
The Inner Elixir (Nei Tan): Chinese physiological alchemy. *Changing perspectives in the history of science*, Nikulaus Teich and Robert Young (eds.). London, Heinnemann, Chap. V, P. 68-84, 1972.

867 LU, Gwei-djen and NEEDHAM, Joseph
China and the origin of examinations in medicine. *Royal Society of Medicine, Proceedings*, Vol. 56, P. 63-70, 1963.
History of medical education.

868 LU, Gwei-djen and NEEDHAM, Joseph
A contribution to the history of Chinese dietetics. *ISIS*, 42(127), 13-20, 1951.
Empirical knowledge of diet, especially in relation to certain deficiency diseases, developed in ancient Chinese civilization. 'At the same time it was only by the analytical methods of Western science that the relations of the food-constituents to health and disease could be incorporated into a logical system.'

869 LU, Gwei-djen and NEEDHAM, Joseph
Medieval preparations of urinary steroid hormone. *Medical History*, 8(2), 101-21, 1964.
Knowledge of the steroid sex hormones is an achievement of modern science, but was also accomplished by the Chinese iatro-chemists between the tenth and sixteenth centuries. Guided by theories of traditional Chinese type, and using urine as their starting point, they succeeded in preparing mixtures of androgens and oestrogens in relatively purified form and employing them in medicine.

870 LU, Gwei-djen and NEEDHAM, Joseph
Records of diseases in ancient China. *Diseases in Antiquity*, Don Brothwell and A.T. Sandison (eds.). Springfield, Charles C. Thomas, Chap. 17, P. 222-37, 1967.
'. . . the study of the written records of ancient China from the middle of the first millenium B.C. down to the beginning of our era, shows that they have preserved a veritable mass of information concerning the diseases prevalent in those times . . . .

871 LU, Gwei-djen, SALAMAN, Raphael A. and NEEDHAM, Joseph
The wheelwright's art in ancient China; I--The invention of 'dishing'; II--Scenes in the workshop. *Physis*, 1(2), 103-26, 1959;

1(3), 196-214, 1959.
Investigation of the methods used in ancient China for construction of wheels for carriages and chariots. 'The chief outcome of the study is the demonstration that the technique of 'dishing', i.e. building the wheel in shape of a flat cone, with a view to the greater strength and convenience resulting, far from being a great achievement of European technology in the +16th century, was fully known and practised from the -4th century onwards in China. The conclusions of the textual study are supported by new archaeological findings.' Part II presents two Han reliefs of the + or -1st century depicting the wheelwright's workshop, and discusses their probable interpretation.

872 MIYASITA, Saburo
A link in the Westward transmission of Chinese anatomy in the later Middle Ages. *ISIS*, Vol. 58, P. 486-90, 1967.
The Ts'un hsin huan chung t'u by Yang Chieh (1113) was transmitted to Japan and Western Asia, and its influence can be seen in the works of the Japanese monk-physician Shozen Kajiwara (1304, 1315) and in the Persian Treasure of the Ilkhan on the Sciences of Cathay (c. 1313-1314). Influence of Chinese ideas on the development of European anatomy as yet unknown.

873 MURPHEY, Rhoads,
*The non-development of science in Traditional China*, papers on China from the Regional Studies Seminar, 1947, unpublished mimeo, P. 1-30. Harvard University, Committee on International and Regional Studies.
'Humanistic interpretation' of natural phenomena left the Chinese 'both undesirous and incapable of developing the abstract physical theory which underlies Western science. The stable and conservative Chinese social order was . . . less important than . . . the basic Chinese ideological pattern.' 'Selected list of indigenous Chinese findings in empirical science,' representing China's major inventions--noting the absence of applied techniques and machines which might have followed from the discoveries. Factors working against the development of science: accidental or environmental (including isolation; nature of the written language); institutional (including structure of Chinese society; non-development of urban, commercial bourgeoisie; stability of the agricultural, bureaucratic state; traditionalism, conservatism, agrarian mentality; large population); ideological factors (including preoccupation with human and social values; pragmatism; view of truth as self-evident; absence of universal law or logic; concept of harmony between man and nature; anti-rational tradition; lack of concept of change or progress; nature of the education system.)

874 NAKAYAMA, Shigeru
Characteristics of Chinese astrology. *ISIS*, Vol. 57, Pt. 4, No. 190, P. 442-454, 1966.
Empirical and theoretical development of Chinese astrology.

875 NAKAYAMA, Shigeru
Kyoto Group of the History of Chinese Science. *Japanese Studies in the History of Science*, No. 9, P. 1-4, 1970.
Description of the work of the Group, including tables of contents of four volumes on Chinese science and technology by historical period.

876 *NATURE*
East and West in science, 169(4306), 774-6, 1952.
Includes review of a paper presented by Dr. J. Needham to a symposium of the British Society for the History of Science, 17 April 1952, on 'Transmission of ideas in the mechanical sciences between China and Europe'.

877 *NATURE*
Science and society in Ancient China, 161(4087), 305, 1948.
Report of the Conway Memorial Lecture by Dr. Joseph Needham, 12 May 1947 (entry no. 907).

878 NEEDHAM, Joseph
Aeronautics in Ancient China, adapted from 'Science and civilization in China'. *Shell Aviation News*, No. 279, P. 2-7, 1961.
The precursors of the aeroplane propeller existed in China. The development of the science of modern aerodynamics in the nineteenth century depended upon the kite, which is Chinese in origin.

879 NEEDHAM, Joseph
Astronomy in Classical China. *Quarterly Journal of the Royal Astronomical Society*, Vol. 3, P. 87, 1962. Also in *Clerks and craftsmen in China and the West*. Cambridge, Cambridge University Press, P. 1-13, 1970.
Before the 17th century, 'Chinese astronomy as a system was quite different from that of the West and its emphasis and understandings.' Instrumentation and apparatus.

880 NEEDHAM, Joseph
Central Asia and the history of science and technology. *Journal of the Royal Central Asian Society*, 36(2), 135-45, 1949. Also in *Clerks and craftsmen in China and the West*. Cambridge, Cambridge University Press, P. 30-39, 1970.
Transmission of scientific and technological discoveries through Sinkiang, Tibet and bordering countries.

881 NEEDHAM, Joseph
China, Europe, and the seas between. *Clerks and craftsmen in China and the West*. Cambridge, Cambridge University Press, P. 40-70, 1970.
Navigation techniques used by ancient Chinese mariners.

882 NEEDHAM, Joseph
China and the invention of the pound-lock. *Transactions of the Newcomen Society*, Vol. 36, P. 85-107, 1963/4.
'The object of the present contribution is to elucidate the background of the invention of the pound-lock by the engineers of mediaeval China . . . .' 'The first recognisable pound-lock

built in China dates from the decade following A.D. 980, the first in Europe from that following A.D. 1370 . . . .' As smaller vessels replaced larger in the northward traffic on canals during the Ming dynasty, the pound-locks 'fell into decay one by one and were not replaced.' Speculation that Chinese engineers were employed to rehabilitate irrigation and water-conservancy works in Mesopotamia, following Mongol conquest and destruction in the 13th century A.D.

883 NEEDHAM, Joseph
China's philosophical and scientific traditions. *Cambridge Opinion*, Vol. 36, P. 11-16, 1963.
Brief discussion of Confucianism, Taoism and Neo-Confucianism, and their relative positions in traditional Chinese society; Chinese culture attained 'sagely synthesis', while a fundamentally unreconciled dualism between Greek atomism and Hebrew spiritualism lay at the root of European culture. The Western belief that China was a stagnant civilization is based on three paradoxes: the belief that there was never any science or technology in China; the bureaucratic ethos helped science and the useful application of natural knowledge in China between the +2nd and +15th centuries; China's slow and steady progress was overtaken by the exponential growth of modern science after the Renaissance.

884 NEEDHAM, Joseph
The Chinese contribution to the development of the mariner's compass. *Clerks and craftsmen in China and the West*. Cambridge, Cambridge University Press, P. 239-49, 1970.
Ancient Mediterranean and East Asian cultures both were aware of the attractive property of lodestone, but the directive properties were known in China a century before noted in Europe.

885 NEEDHAM, Joseph
The Chinese contribution to science and technology. *Reflections on our age*, Hardman, D. and Spender S. (eds.). London, Allan Wingate Ltd., 1948. Also in *Clerks and craftsmen in China and the West*. Cambridge, Cambridge University Press, P. 71-82, 1970.
Chinese scientific and technological achievements: why did their science and technology always remain primarily empirical? Why was there no indigenous industrial revolution in China? 'China was fundamentally an irrigation-agricultural civilization, as contrasted with the pastoral-navigational civilization of Europe;' in Chinese 'bureaucratic feudal' society, the merchant class could not rise to power, and thus finance research for new forms of production and trade.

886 NEEDHAM, Joseph
The Chinese contributions to vessel control. *Clerks and craftsmen in China and the West*. Cambridge, Cambridge University Press, P. 250-62, 1970.
Development and use of steering mechanisms in medieval Chinese nautical technology.

887 NEEDHAM, Joseph
Chinese priorities in cast iron metallurgy. *Technology and Culture*, Vol. 5, P. 398-404, 1964.
Comparison of certain technological developments in China and Europe.

888 NEEDHAM, Joseph
*Classical Chinese contributions to mechanical engineering*. Earl Grey Memorial Lecture, 28 Feb 1961, King's College, Newcastle upon Tyne. Also in *Clerks and craftsmen in China and the West*. Cambridge, Cambridge University Press, P. 113-35, 1970.
The water-wheel and the water-mill as labor-saving inventions; use of water-power for metallurgical blowing engines and their relationship to reciprocating steam engine. Chinese paddle-wheel boats. Time-keeping; the water-wheel link-work escapement. Adoption of techniques known and used in China for centuries in 14th century Europe. Enthusiasm in modern China for science and the history of science.

889 NEEDHAM, Joseph
*Clerks and craftsmen in China and the West*. Cambridge, Cambridge University Press, 1970.
Contents: Astronomy in Classical China; the unity of science: Asia's indispensable contribution; Central Asia and the history of science and technology; China, Europe, and the seas between; the Chinese contribution to science and technology; the translation of old Chinese scientific and technical texts; the earliest snow crystal observations; iron and steel production in Ancient and Medieval China; classical Chinese contributions to mechanical engineering; the pre-natal history of the steam-engine; the missing link in horological history: a Chinese contribution; the Chinese contribution to the development of the mariner's compass; the Chinese contributions to vessel control; medicine and Chinese culture; proto-endocrinology in Medieval China; elixir poisoning in Medieval China; hygiene and preventative medicine in Ancient China; China and the origin of qualifying examinations in medicine; the roles of Europe and China in the evolution of oecumenical science.

890 NEEDHAM, Joseph
The development of iron and steel technology in China. *Second Biennial Dickinson Memorial Lecture to the Newcomen Society 1956*. Cambridge, W. Heffer and Sons for the Newcomen Society, 1964.
'Epitome' as iron and steel production in Ancient and Medieval China in *Clerks and craftsmen in China and the West*. Cambridge, Cambridge University Press, P. 107-12, 1970.
Technological development from the -6th century on.

891 NEEDHAM, Joseph
*The dialogue of Europe and Asia*. Presidential address, British-China Friendship Association, Oct 1955. Also *The Glass Curtain between Asia and Europe*, Iyer, Raghavar (ed.). London, Oxford University Press, Chap. XVIII, P. 279-96, 1965.
'The real universal factors are modern science and modern technology, together with the philosophies which made them

possible.' Europeans must acknowledge oriental antecedants of modern science and share the fruits of modern science.

892 NEEDHAM, Joseph
*The Grand Titration: science and society in East and West.* London, George Allen and Unwin Ltd., 1969.
Contents: Poverties and triumphs of the Chinese scientific tradition; science and China's influence on the world; on science and social change; science and society in Ancient China; thoughts on the social relations of science and technology in China; science and society in East and West; time and Eastern man; human law and the laws of nature.

893 NEEDHAM, Joseph
Hand and brain in China. *China Now,* No. 5, P. 5-7, Sep/Oct 1970; No. 6, P. 5-7, Nov 1970; No. 7, P. 5-7, Dec 1970; No. 8, P. 4-7, Jan 1971.
Text of a lecture at the Annual General Meeting of the Society for Anglo-Chinese Understanding, 16 May 1970. Current Chinese scientific developments related to past inventions and discoveries. Pt. 1--China's exploding technology. 2--From seismograph to dynamo. 3--Wooden tankers and one-man workshops. 4--The alchemy of human nature.

894 NEEDHAM, Joseph
How the Chinese invented the mechanical clock. *New Scientist,* 4(108), 1481-1483, 1958.
The principle which unites the clepsydra with the mechanical clock in a direct line of evolution was embodied in the huge clock which Su Sung designed and built between 1088 and 1092, powered by water and using an escapement, and described in Su Sung's monograph of 1090. This description tallies with similar devices incompletely described in Indian, Arabic and Hispano-Moorish texts. Failure to evolve a term for purely time-telling clockwork (after I-Hsing began it in the eighth century) led to the erroneous impression that the Jesuits introduced clockwork to China.

895 NEEDHAM, Joseph
*Human law and the laws of nature.* Hobhouse Lecture, Bedford College, London 1951. Also *Journal of the history of ideas,* Vol. 12, P. 3, 194, 1951. Also in *The Grand Titration.* London, George Allen & Unwin Ltd., P. 299-331, 1969.
Modern science and the philosophy or organism, fortified by new understanding of cosmic, biological, and social evolution, have come back to the Chinese world-view of the harmonious cooperation of all beings, arising from the fact that they were all parts of a hierarchy of wholes forming a cosmic pattern . . . .

896 NEEDHAM, Joseph
Mathematics and science in China and the West. *Science and Society,* Vol. 20, P. 320-43, 1956.
'The aim of the following pages is to open a discussion on the relations of mathematics to natural science in two diverse

civilizations, Europe and China . . . . Exactly what were the relations of mathematics to science in ancient and medieval China? What was it that happened in Renaissance Europe when mathematics and science joined in a combination qualitatively new and destined to transform the world?'

897 NEEDHAM, Joseph
The missing link in horological history: a Chinese contribution. *Proceedings of the Royal Society,* Vol. 250, P. 147, 1959. Also in *Clerks and craftsmen in China and the West.* Cambridge, Cambridge University Press, P. 203-38, 1970.
The escapement was developed in China and was first applied to the water-wheel. In Europe, such artisans as millwrights were one of the important roots of Renaissance science, but since they were 'at least as eminent' in skill and ingenuity in China, the presence of artisans was evidently not in itself enough.

898 NEEDHAM, Joseph
The Peking Observatory in A.D. 1280 and the development of the equatorial mounting. *Vistas in Astronomy,* Arthur Beer (ed.). London, Pergamon, Vol. 1, P. 67-83, 1955.
The 13th century scientist and engineer, Kuo Shou-Ching, modified mediaeval Arabic and European astronomical instruments brought to China in 1297 by the Persian astronomer, Jamal al-Din, sent by the Ilkhan to confer with the astronomers of the observatory at Peking.

899 NEEDHAM, Joseph
Poverties and triumphs of the Chinese scientific tradition. *Scientific Change,* A.C. Crombie (ed.). London, Heinemann, P. 117-53, 1963. Also in *The Grand Titration.* London, George Allen & Unwin Ltd., P. 14-54, 1969.
Analysis of the social and economic structures of Eastern and Western cultures explains why modern science rose only in the West at the time of Galileo and why, between the -2nd century and +15th century, East Asian culture applied human knowledge of nature to useful purposes more efficiently than the European West. Science and technology in traditional China. Contrasts between China and the West. Social position of scientists and engineers in traditional China. Feudal-bureaucratic society. Old World origins of the new science.

900 NEEDHAM, Joseph
The pre-natal history of the steam engine. *Transactions of the Newcomen Society,* Vol. 35, 1962-63. Also in *Clerks and craftsmen in China and the West.* Cambridge, Cambridge University Press, P. 136-202, 1970.
'. . . the entire morphology (and some of the physiology) of the reciprocating steam-engine of the early nineteenth century was prefigured in Asian, especially Chinese, machinery, widely used at the beginning of the thirteenth.'

901 NEEDHAM, Joseph
Prospection Geobotanique en Chine Medievale. *Journal d'Agricultur Tropicale et de Botanique Applique,* 1(5/6), 143-7, 1954.

The methods which were used by the ancient miners in the search for minerals and ores was principally based, without doubt, on traditional geology, observation of the topography of the territory, the direction of strata and knowledge of rocks. It suggests the work in sixteenth century Europe.

902 NEEDHAM, Joseph
*Relations between China and the West in the history of science and technology.* Actes du Septieme Congres International d'Histoire des Sciences, Jerusalem 1953.
Discussion of the interchanges of science and technology between China and Europe through history. '. . . it is . . . well to make a distinction between scientific theories and observations as such, on the one hand, and technological inventions on the other . . . . Was it not the inventions of immediate practical use which tended to travel, rather than the scientific and pre-scientific observations, speculations and theories? . . . . Broadly speaking, Chinese science, for two millenia before the coming of the Jesuits, and in spite of opportunities of intellectual intercourse much greater than has often been pictured, had very little in common with that of the West . . . . But Chinese technological inventions poured into Europe in a continuous stream during the first thirteen centuries of the Christian era, just as later on the technological current flowed the other way.'

903 NEEDHAM, Joseph
The roles of Europe and China in the evolution of oecumenical science. *Advancement of Science,* Vol. 24, Sep 1967. Also in *Clerks and craftsmen in China and the West.* Cambridge, Cambridge University Press, P. 396-418, 1970.
The mediaeval sciences of West and East were subsumed in modern science. 'When in history did a particular science in its Western form fuse with its Chinese form so that all ethnic characteristics melted into the universality of modern science?'; 'at what point in history did the Western form decisively overtake the Chinese form?'

904 NEEDHAM, Joseph
Science and China's influence on the world. *The Legacy of China,* Dawson, Raymond S. (ed.). Oxford, Clarendon Press, P. 234-308, 1964. Also in *The Grand Titration.* London, George Allen & Unwin Ltd., P. 55-122, 1969. Extracts in *Eastern Horizon,* 5(2), 21-32 (Pt. I), 1966; 5(3), 10-20 (Pt. II), 1966.
China dominated technological influences before and during the Renaissance: efficient equine harness, technology of iron and steel, gunpowder and paper, mechanical clock, basic engineering devices. Chinese scientific discoveries: immunization technique; cosmology; undulatory theory; seismograph; biological and pathological classification systems; anatomy; pharmacology. Technical inventions: paddle-wheel boat; iron-chain suspension bridge; differential gear; steel-making methods. Chinese practical inventions were transmitted to the West, but Chinese naturalistic theory, recorded experimentation and measurement was hemmed in within the boundaries of the ideographic language. Chinese inventions were not merely technological achievements, e.g.

invention of gunpowder was based on Taoist alchemy. Effect of certain inventions on Chinese social structure.

905 NEEDHAM, Joseph
*Science and civilization in China.* Vol. 5--Chemistry and industrial chemistry (projected). Vol. 6--Biology, agriculture and medicine (projected).

906 NEEDHAM, Joseph
*Science and civilization in China.* Vol. 7--The social background (projected).
'Why did modern science . . . develop round the shores of the Mediterranean and the Atlantic, and not in China or any other part of Asia? This is the question to which the fourth part is devoted. Its consideration involves an examination of the concrete environmental factors of geography, hydrology, and the social and economic system which was conditioned by them, though it cannot leave out of account questions of intellectual climate and social customs. At the end of this part, the problem of parallel types of civilization is touched upon.' Contents: Retrospective survey of the characteristics of Chinese science; Geographical factors; Social and economic factors; Philosophical and ideological factors; General conclusions.

907 NEEDHAM, Joseph
Science and society in Ancient China. *Conway Memorial Lecture, 12 May 1947.* London, Watts, 1947. Also *Mainstream,* 13(7), 7, 1960. Also in *The Grand Titration.* London, George Allen & Unwin Ltd., P. 154-76, 1969.
Origins of the first Chinese society; Chinese philosophy. Confucian ethical rationalism was antagonistic to the development of science, whereas Taoist empirical mysticism was in favour of it. Ancient Chinese feudalism is analogous with the European Bronze Age/Iron Age: state of military technology, lack of large-scale slavery. This was replaced by 'Asiatic bureaucratic' society, which prevented the rise of the merchant class to power and thus inhibited the rise of modern science and technology in China.

908 NEEDHAM, Joseph
Science and society in China and the West. *Science Progress,* Vol. 52, P. 50-65, Jan 1964. Also Glories and defects of the Chinese scientific and technical traditions in Welskopf, E.C. (ed.) *Neue Beiträge zur Geschichte der alten Welt,* Band I: Alter Orient und Griechenland. Berlin, Akademie-Verlag, P. 87-109, 1964.
'My object will be to describe some of the outstanding contrasts between the Chinese and European traditions in the natural sciences, pure and applied, then to say something about the position of scientists and engineers in classical Chinese society, and lastly to discuss certain aspects of science in relation to philosophy, religion, law, language and the concrete circumstances of production and exchange of commodities.' '. . . if parallel social and economic changes (to Europe) had taken place in Chinese society then some form of modern science would have arisen there.'

909 NEEDHAM, Joseph
Science and society in East and West. *Science and Society*, 28(4), 385-408, 1964. Also *The science of science*, Mackay, A. and Goldsmith M. (eds.). London, Souvenir Press, P. 127-49, 1964. Also *Centaurus*, Vol. 10, P. 174-97, 1964. Also in *The Grand Titration*. London, George Allen & Unwin Ltd., P. 190-217, 1969.
The answer to the questions of why modern science had not developed in Chinese civilization, but only in Europe, and why Chinese civilization between the -1st century and +15th century was more efficient than occidental in applying human knowledge to practical human needs . . . lies in the social, intellectual, and economic structures of the different civilizations.

910 NEEDHAM, Joseph
Thoughts on the social relations of science and technology in China. *Centaurus*, Vol. 3. P. 40-48, 1953. Also in *The Grand Titration*. London, George Allen & Unwin Ltd., P. 177-89, 1969.
'There cannot be much doubt . . . that the failure of the rise of the merchant class to power in the state lies at the basis of the inhibition of the use of modern science in Chinese society.' Chinese bureaucratic society was inferior to the society of the European Renaissance in technical creativity but more successful than European feudalism or the slave-owning Hellenistic society.

911 NEEDHAM, Joseph
Time and Eastern man. *Henry Myers Lecture*, Royal Anthropological Institute, 1964. Also *The voices of time*, Fraser, J.T. (ed.). New York, 1966. Also in *The Grand Titration*. London, George Allen & Unwin Ltd., P. 218-98, 1969.
'If Chinese civilization did not spontaneously develop modern nature science as Western Europe did . . . it was nothing to do with her attitude towards time.'

912 NEEDHAM, Joseph
Understanding past is key to future. *Far East Trade and Development*, 20(10), 977-81, 1965.
Chinese science and technology generally more advanced than in Europe until overtaken by the 'scientific revolution' in Europe in the 15th century. But modern science is universal, not Europocentric, but this does not imply that everything European is also universal. Europe and America must transmit modern science and technology to Asia and Africa, and learn from them.

913 NEEDHAM, Joseph
The unity of science: Asia's indispensable contribution. *Clerks and craftsmen in China and the West*. Cambridge, Cambridge University Press, P. 14-29, 1970.
Arabic studies of East Asian science were not translated into Latin by medieval Europeans. The barrier between East Asian science and the 'Franks and Latins' did not extend to technology: 'technical inventions show a slow but massive infiltration from east to west throughout the Christian era.' Neither Indian nor Western influences on Chinese science were lasting. Comparison of technological development in China and Europe.

914 NEEDHAM, Joseph
*Wheels and gear-wheels in Ancient China*. Barcelona, Actes du IXe Congres International d'Histoire des Sciences, Vol. 2, P. 222-25, 1959.
Textual and archaeological evidence supports the view that 'dishing' was practised in the building of wheels in China from the -4th century on. Wooden wheels on Han vehicles are related to wooden gear-wheels and smaller, bronze toothed wheels. Parallel development of roller bearings in China and in Hellenistic Greece. 'The conclusion seems to be presenting itself more and more unescapably, that both the Chinese and the Hellenistic Greeks built independently on some very simple origins bequeathed to both of them by Mesopotamian culture.'

915 NEEDHAM, Joseph
*Within the four seas: the dialogue of East and West*. London, George Allen & Unwin Ltd., 1969.
1--The dialogue of East and West. 2--The past in China's present. 3--The Chinese contribution to scientific humanism. 19--Psychology and scientific thought in East and West. 21--Science and religion in the light of Asia. Traditional Chinese science and the non-development of modern science in China.

916 NEEDHAM, Joseph, BEER, Arthur and HO Ping-Yü
'Spiked' comets in Ancient China. *The Observatory*, 77(899), 137-8, 1957.
References to comets with two tails or 'horns' in Chinese dynastic histories.

917 NEEDHAM, Joseph and LU, Gwei-djen
The earliest snow crystal observations. *Weather*, 16(10), 319-27, 1961. Also in *Clerks and craftsmen in China and the West*. Cambridge, Cambridge University Press, P. 98-106, 1970.
Chinese observation of snow-flake forms dates from 2nd century B.C., antedating first European observations by more than a millenium. 'Perhaps one might be inclined to see in this parallel development of knowledge of snowflakes in East and West an epitome of the difference between the European and Chinese social environments. The Chinese . . . very early observation . . . was allowed to become a commonplace and relatively little development occurred through the centuries . . . .'

918 NEEDHAM, Joseph
Efficient equine harness: the Chinese inventions. *Physis*, Vol. 2, P. 121, 1960. A further note on efficient equine harness: the Chinese inventions. *Physis*, Vol. 7, P. 70-74, 1965.
The prior paper infers the invention of collar-harness for equine animals from representations of horses and vehicles in the fresco-paintings at the cave-temples near Tunhuang, ca. +5th century. In the latter paper, the authors' discovery in 1964 of 'a very clear carving of collar-harness' in the Yünkang cave-temples (ca. +477), is reported. See also *Mediaeval technology and social change*, Lynn White (ed.). Oxford University Press, P. 157, 1962.

919 NEEDHAM, Joseph and LU Gwei-djen
The esculentist movement in mediaeval Chinese botany; studies on wild (emergency) food plants. *Archives Internationales d'Histoire des Sciences*, 21(84/85), 225-48, 1968.
The botanist extended 'the frontiers of knowledge to certify what plants were safe and wholesome, what dangerous and evil, for the welfare of the people' in times of food scarcity and famine. 'Thus it was that in the +14th century the Chinese botanists became involved with iatro-chemical procedures and what we should call biochemistry today; evidently they experimented carefully and widely.' 'We do not know of any parallel in European, Arabic or Indian mediaeval civilization.' 'It seems that it was not until the 18th century that similar interests began to prevail in Europe.' '. . . To anyone who designates the Ming period as one of scientific decline or 'stagnancy' in Chinese scientific endeavour, one can answer that it saw almost the whole on this great and unprecedented effort to extend the realm of botany from plants believed to be of pharmaceutical value to include all those which were useful for the diet of man.'

920 NEEDHAM, Joseph and LU Gwei-djen
Hygiene and preventive medicine in Ancient China. *Health Education Journal*, Sep 1959. Also in *Clerks and craftsmen in China and the West*. Cambridge, Cambridge University Press, P. 340-78, 1970.
Historical observations on the concepts of hygiene and preventive medicine in ancient China down to the end of the Han.

921 NEEDHAM, Joseph with LU Gwei-djen
Medicine and Chinese culture. *Clerks and craftsmen in China and the West*. Cambridge, Cambridge University Press, P. 263-93, 1970.
General position of medicine in traditional Chinese society; influence of philosophical and religious doctrines upon Chinese medicine; effects of the transition from traditional society to Marxist society.

922 NEEDHAM, Joseph and LU Gwei-djen
The optick artists of Chiangsu. *Proceedings of the Royal Microscopical Society*, 1(2), 113-38, 1966 (abstract p. 59-60).
Brief biographies of Po Yü (b. ca. 1610) and Su Yün-Chhiu (ca. 1630-1663) reconstructed from local gazeteers. Po Yü may have independently invented the telescope--but this cannot be determined without knowing the extent of Jesuit influence in Suchow. Optical devices of Sun Yün-Chhiu include telescope, microscope, various spectacles. 'It is truly remarkable to find Chinese scientific and technical men following so closely upon the heels of the pioneers of optical apparatus in Western Europe . . .' 'The story . . . illustrates the speed with which Chinese mathematics, astronomy and physics could fuse with these sciences in their Western forms to give modern universal world science.'

923 NEEDHAM, Joseph with LU Gwei-djen
Proto-endocrinology in Medieval China. *Japanese studies in the history of science*, Vol. 5, P. 150, 1966. Also in *Clerks and craftsmen in China and the West*. Cambridge, Cambridge University

Press, P. 294-315, 1970.
'A brief account of the pre-history of endocrinology as it can be found in ancient and medieval China.'

924 NEEDHAM, Joseph and LU, Gwei-djen
Sex hormones in the Middle Ages. *Endeavour*, 27(102), 130-2, 1968.
'Thus there can be little doubt that between the eleventh and the seventeenth century A.D. the Chinese iatro-chemists were producing preparations of androgens and oestrogens which were probably quite effective in the quasi-empirical therapy of the time.'

925 NEEDHAM, Joseph and ROBINSON, Kenneth
Ondes et particules dans la pensee scientifique Chinoise. *Sciences*, 1(4), 65-78, (1960?).
Ancient and medieval Chinese thought was organic and continuous, a cosmos of waves rather than of particles. The origins of modern physics can thus be traced back to the Chinese and to the Stoics.

926 NEEDHAM, Joseph with LING, Wang
*Science and civilization in China*. Cambridge, Cambridge University Press, 1954. Vol. 1: Introductory Orientations.
'The introductory part consists of brief accounts of (a) the geographical background, (b) the history of China, (c) the special characteristics of the Chinese language, and (d) the opportunities which the centuries afforded to culture contact whereby the passage of scientific ideas or technological processes to and from East Asia may have been made possible or even facilitated.' Contents: 4--Geographical introduction: general survey of Chinese topography; the geotectonics of China; human geography of the natural provinces. 5--Historical introduction. The pre-Imperial phase. 6--Historical introduction. The empire of all under heaven. 7--Conditions of travel of scientific ideas and techniques between China and Europe: introduction; the originality of Chinese culture; rumours of Chinese culture in the classical West; the continuity of Chinese with Western civilization; the development of overland trade routes; the development of the maritime trade-routes; the old silk road; Chinese-western cultural and scientific contacts as recorded by Chinese historians; Chinese-Indian cultural and scientific contacts; Chinese-Arab cultural and scientific contacts.

927 NEEDHAM, Joseph
*Science and civilization in China*. Cambridge, Cambridge University Press, 1956. Vol. 2: History of scientific thought.
'The second part takes up the question of the origin and development of scientific thought in Chinese philosophy . . . apart from the vision of the Taoists, there runs throughout Chinese history a current of rational naturalism and of enlightened scepticism . . .' Contents: 9--The Ju Chia (Confucian) and Confucianism. 10--The Tao Chia (Taoists) and Taoism. 11--The Mo Chia (Mohists) and the Ming Chia (Logicians). 12--The Fa Chia (Legalists). 13--The fundamental ideas of Chinese science. 14--The pseudo-sciences and the sceptical tradition, 15--Buddhist thought. 16--Chin and Thang Taoists, and Sung Neo-Confucians. 17--Sung and Ming idealists, and the last great figures of indigenous

naturalism. 18--Human law and the laws of nature in China and the West.

928 NEEDHAM, Joseph with LING, Wang
*Science and civilization in China*. Cambridge, Cambridge University Press. Vol. 3 Mathematics and the sciences of the heavens and the earth (1959). Vol. 4. Physics and physical technology: Pt. I--Physics (1962); Pt. II--Mechanical engineering (1965); Pt. III--Civil engineering and nautics (1971).
'In the third part, which deals with the sciences, pure and applied, in due order, we attempt to answer the question, what exactly did the Chinese contribute to science, pure and applied, through the historical centuries . . . . The Chinese pre-sciences, both ancient and medieval, show the clearest development of experimental and observational inductive science, involving manual operations, though they were always interpreted by theories and hypotheses of primitive type . . . . Similarly, Chinese technology, both ancient and medieval, led to empirical discoveries and inventions, many of which profoundly affected world history. It is quite clear that the Chinese could plan and carry out useful experiments for the further improvement of techniques, though again always interpreting them by theories of primitive type.'

928a SIVIN, Nathan
Science and civilization in China. *Journal of Asian Studies*, 27(4), 859-63, 1968.
Review of Joseph Needham with Wang Ling, *Science and civilization in China*, Vol. 4, Physics and physical technology; Pt. II, Mechanical engineering. The central thesis of Needham's work is that the crucial determinants of China's scientific achievement--and ultimately of its failure to develop modern science were preponderantly social and economic rather than due to intellectual factors. 'I am convinced that this book is the most important contribution to the advancement of its field in many decades.'

928b SIVIN, Nathan
Science and civilization in China, Vol. 4: Physics and physical technology, Pt. III--Civil engineering and nautics. *Scientific American*, 226(1), 113-18, 1972.
Review of the work by Joseph Needham, in collaboration with Lu Gwei-djen. Needham aims to answer the questions, 'How was China able to attain an early scientific and technological superiority over the West, so that until the Renaissance it had more to give than to receive? And why, in spite of this early lead, did that improbable sequence of events, the scientific revolution, happen in Europe and not in China?' Chinese technology appears to have been quite commensurable with that of Western classical antiquity, indicating a common archaic center of diffusion in Mesopotamia and perhaps in Egypt. Needham argues that transmission of Chinese techniques to Europe after the Dark Ages is more likely than independent reinvention. Brief review of the scheme of the seven volumes of *Science and civilization in China* and themes common in Needham's writings.

928c WHITE, Lynn Jr.
More pieces to the Chinese puzzle. *ISIS*, Vol. 58, P. 248-51, 1967. Review of Joseph Needham with Wang Ling, *Science and civilization in China*, Vol. 4; Pt. II, mechanical engineering. Needham is writing the histories of Chinese science and technology for the first time and integrating these histories with the general development of China and of the rest of the world. Chinese science is of interest in the interpretation of East Asian cultures; Chinese technology, on the other hand, has global significance.

929 NEEDHAM, Joseph, LING Wang and PRICE, Derek J.
Chinese astronomical clockwork. *Nature*, 177(4509), 600-602, 1956. Also in *Actes du VIIIe Congres International d'Histoire des Sciences*. Florence, P. 325 ff. 1956.
A long tradition of astronomical clock-making in China between the 7th and 14th centuries A.D. The Chinese tradition was in the direct line of ancestry of the late medieval European mechanical clocks; the time of transmission was the Crusades.

930 NEEDHAM, Joseph, LING Wang and DE SOLLA PRICE, Derek J.
*Heavenly clockwork: the great astronomical clocks of Medieval China*. London, Cambridge University Press, 1960.
'It is generally allowed that the invention of the mechanical clock was one of the most important turning-points in the history of science and technology . . . . The examination of certain medieval Chinese texts . . . now permits us to establish the existence of a long tradition of astronomical clock-making in China between the seventh and fourteenth centuries A.D., and perhaps even having its origins as early as the second century A.D.'

931 RETI, Ladislao
The double-acting principle in East and West. *Technology and Culture*, 11(2), 178-200, 1970.
History of use of the double-acting principle in technology in China and Europe.

932 SCHRIMPF, Robert
Bibliographie Sommaire des Ouvrages Publies en Chine durant la periode 1950-1960 sur l'Histoire du developpement des sciences et des techniques Chinoises. *Bulletin Ecole francaise d'Extreme Orient*, Vol. 51, P. 615-24, 1963.
Summary bibliography of works published in China during the period 1950-1960 on the history of the development of Chinese science and technology. I--Collected studies of science and technology. II--Mathematical and astronomical sciences. III--Chemical sciences. IV--Techniques for water control.

933 SIVIN, Nathan
*Cosmos and computation in early Chinese mathematical astronomy*. Leiden, E.J. Brill, 1969.
In Han dynasty sources of the first century A.D., the calendrical treatises attempt to derive the fundamental astronomical constants from a yin-yang and five-elements analysis of cycles of change. 'Between the origin of Chinese astronomy and its full

flowering as a mathematical science in the Sui and T'ang (sixth-tenth centuries A.D.), the sense of cosmos almost completely dropped out.' '. . . what killed the conviction that astronomy could be physics as well as mathematics? . . . the blighting hand of bureaucracy? . . . the vested interests of philosophers? . . . or grave contradictions (in the Han astronomers' mathematical procedures)?'

934 SIVIN, Nathan
Chinese alchemy: preliminary studies. *Harvard monographs in the history of science.* Cambridge, Harvard University Press, 1968.
'My fundamental concern is with the history of ideas.' An investigation of the Tan ching yao chueh and its author, Sun Su-mo (ca. 673). I--On the reconstruction of ancient Chinese alchemy; P. 5-11 Priorities in the study of Chinese science; 'The Chinese tradition is certainly science, by any definition not utterly parochial, but except on the level that makes it science, its goals so consistently diverge from ours that most similes become gratuitous.' II--Tan Ching Yao Chueh: the tradition and the book. III--The biography of Sun Ssu-mo: a historiographic inquiry. IV--Tan Ching Yao Chueh: an annotated translation. V--Tan Ching Yao Chueh: critical edition of the text.

935 SMITH, P.J. and NEEDHAM, Joseph
Magnetic declination in mediaeval China. *Nature,* 214(5094), 1213-14, 1967.
Earliest recorded direct observations of the earth's magnetic field were made by the Chinese.

936 STRUBELL, Wolfgang
Über die Gewinnung and Verwendung von Erdöl im alten China. *Erdöl and Kohle-Erdgas-Petrochemie,* 21(7), 435-40, 1968.
Production and application of petroleum in ancient China. The various production methods and the application of petroleum in ancient China are described on the basis of new and old literature.

937 SUN, E-tu Zen
Sericulture and silk textile production in Ch'ing China. *Economic organization in Chinese society,* W.E. Willmott (ed.). Stanford, Stanford University Press, P. 79-108, 1972.
Characteristics of the silk industry, emphasizing traditional methods, including production of raw silk and trade in silk goods. The 'baseline from which modernization was to start': handicraft technology remained on its pre-industrial plateau.

938 SUN, E-tu Zen
Wu Ch'i-chun: profile of a Chinese scholar-technologist. *Technology and Culture,* 6(3), 394-406, 1965.
Biography of Ch'ing official (1789-1847), author of two technical books on mining and plants, which 'provide us with valuable source material on pre-modern Chinese technology, and throw much light on the character of the environment in which Chinese scholar-technologists worked.'

939 SUNG, Ying-hsing
*T'ien-kung K'ai-wu: Chinese technology in the seventeenth century.* Translated and annotated by E-tu Zen Sun and Shiou-chuan Sun. Penns. State University Press, 1966.
Description of traditional Chinese technology in 1637 by Chinese scholar. Chapters include: the growing of grains; clothing materials; dyes; the preparation of grains; salt; sugars; ceramics; casting; boats and carts; hammer-forging; calcination of stone; vegetable oils and fats; paper; the metals; weapons; vermilion and ink; yeasts; pearls and gems.

940 TS-AO, T'ien-ch'in, HO, Ping-yu and NEEDHAM, Joseph
An early mediaeval Chinese alchemical text on aqueous solutions. *Ambix*, 7(3), 1959.
The text of the San-shih-liu Shui Fa (Thirty-six methods for the bringing of solids into aqueous solutions) throws considerable light on the earliest beginnings of the chemistry of inorganic reactions in aqueous medium.

941 WARE, J.R.
*Alchemy, medicine and religion in the China of A.D. 320; the Nei P'ien of Ko Hung.* Cambridge, MIT Press, 1966.

942 YABUUCHI, Kyoshi
*China's science and Japan.* Tokyo, Mainichi Newspaper Co., 1972.
Collection of various articles published in newspapers and magazines by the author in the previous ten years, rewritten for publication in the present volume. Chap. I--Japanese science and technology. II--Chinese scientific thought--Eastern and Western astronomy; imports of Western science in the Ming and Ch'ing; Western and Chinese calendrical reckoning; traditional medicine in China; historical heritage and the new China. III--Personages in the history of Chinese science: Mo-tze; Mei Wen-ting; the geologist, Ting Wen-chiang. IV--Miscellany.

943 YABUUCHI, Kiyoshi (Yabuuti, Kiyosi)
Sciences in China from the fourth to the end of the twelfth century. *Journal of World History (Cahiers d'Histoire Mondiale)*, 4(2), 330-47, 1958. Also in *The evolution of science; readings in the history of mankind*, Metraux, Guy S. and Francois Crouzet, (eds.). New York, Mentor Books, P. 108-127, 1963.
I--The period of the Northern and Southern dynasties: mathematics and astronomy; medical science and pharmacology; agricultural techniques. II--Sui-T'ang period: astronomy and mathematics; medical science and pharmacology; printing and paper-making. III--Wu-Tai and Sung period: astronomy, cartography, and mathematics; medical science and pharmacology; gunpowder and magnetism.

944 YONG, Lam-lay
The geometrical basis of the ancient Chinese square-root method. *ISIS*, 61(206), Pt. 1, P. 92-102, 1970.
The Chinese, as early as the Han dynasty, had already evolved a basic method for root extraction . . . . The procedures are found in the Chiu chang suan shu. This paper translates relevant passages from the Hsiang chieh chiu Chang suan fa (1261) which illustrate this procedure very lucidly.

# Author Indices

The items catalogued in this bibliography are all 'secondary' in that none originate from the People's Republic of China. However, some of these materials are 'primary' in the sense that they record the observations of foreign visitors to China, who describe Chinese scientific and educational institutions, industrial and agricultural technologies, etc. from their own direct experience. Others include substantial translations from original Chinese documents along with commentary, interpretation and analysis and thus are of a mixed primary and secondary nature. Tertiary materials include 'commentary on the commentary', methodological studies, and anthologies of secondary articles about Chinese science and technology.

The author indices are in three parts according to these three categories of literature. In some cases, the assignment of a particular item to one or another of these three classes had necessarily to be somewhat arbitrary. However, this arrangement is intended to help the reader identify and select the literature appropriate to his own purposes.

The author's name, title of the publication in English, date of publication, and entry number in this bibliography are supplied in the index. The language of items not in English or not available in English translation, is indicated by the following code: Da, Danish; Du, Dutch; F, French; G, German; I, Italian; J, Japanese; Sp, Spanish; Sw, Swedish.

## Author Index I. *Primary Material.*

## Author Index II. *Secondary Material.*

## Author Index III. *Tertiary Material.*

AARKROG, A. and J. LIPPERT
Comparison of relative radio-nuclide ratios in debris from the third and fifth Chinese nuclear test explosions (1967) 307

ADOLPH, William H.
Chemistry in China (1946) 766

ALLARD, BLAIN and RIVIERE
The French Steel Mission in China (1965) F 599

ALLEY, P.J.
Some engineering universities in China (1966) 402

AMERICAN MEDICAL NEWS
Inside look at Chinese medicine (1971) 685

BAIRD, John
Public health in China (1956) 686

BARRIENTOS, Celso, H.W. FEELY and D. KATZMAN 322

BERNAL, J.D.
A scientist in China, Pt. I; Universities and colleges, Pt. II (1955) 121

BEST, J.B.
Impressions of Chinese medical services (1967) 687

BLAIN, ALLARD and RIVIERE 599

BOWERS, John Z.
Medicine in Mainland China; Red and Rural (1970) 688

BURHOP, R.H.S.
Physics in Modern China (1956) 262

CHAMBERS, Sir Paul
Trading with China (1966) 657

CHIANG Kai-shek
The way and spirit of science (1943) 778

CHOUARD, Pierre
Scientific research on the morrow of the Cultural Revolution (1972) F 228

CHRISTIANSEN, Wilbur N.
A foreign scientist in the Chinese Cultural Revolution (1970) 229

CHRISTIANSEN, W.N.
A radio astronomer visits China (1964) 269

CHRISTIANSEN, W.N.
Science and scientists in China today (1967) 230

DALYELL, Tam
Chemical industry in China today (1972) 523

DALYELL, T.
China--a market for British know-how now (1971) 658

DALYELL, T.
Scottish mission to China stresses stamina to sustain technical cooperation (1972) 659

DATAMATION
China has some computer building (1972) 271

DAVIS, Chandler
A mathematical visit to China (1971) 258

DIMOND, E. Grey
Medical education and care in People's Republic of China (1971) 702

DOVER, Cedric
The use of biology in China (1956) 279

EISENBERG, A. et al
Fresh fall-out in Israel from the second Chinese nuclear detonation (1966) 319

ELLIOTT, K.A.C.
Observations on medical science and education in the People's Republic of China (1965) 704

ERRERA, M.
Molecular biology in China (1965) 280

FAUNDRES, Anibal and Tapani LUUKKAINEN
Health and family planning services in the Chinese People's Republic (1972) 738

FEELY, H.W., Celso BARRIENTOS and D. KATZMAN
Radioactive debris injected into the stratosphere by the Chinese nuclear weapon test of May 9 1966 (1966) 322

FRASER, Stewart
Chinese Communist education: records of the first decade (1965) 420

GALLAGHER, Louis J.
China in the 16th century: the journals of Matthew Ricci (1942) 845

GALSTON, Arthur W.
Attitudes on acupuncture (1972) 707

GALSTON, A.W.
No grades, no tests (1972) 281

GALSTON, A.W.
The university in China (1972) 422

GALSTON, A.W. and Ethan SIGNER 190

GREGORY, J.W.
The scientific renaissance in China (1924) 785

HAMBRAEUS, Gunnar
Experience from China just now (1972) Sw 68

HAMBRAEUS, G.
Science and technology in China after the Cultural Revolution (1972) 144

HIKOTARO, Ando
Record of a visit to Peking University (1963) J 425

HINSHELWOOD, Sir Cyril
A visit to China (1959) 147

HO, Ping-yu and Joseph NEEDHAM
Theories of categories in Early Mediaeval Chinese alchemy (1959) 852

HO, Ping-yu, T'ien-ch'in TS'AO and Joseph NEEDHAM 940

HORN, J.S.
Away with all pests . . . an English surgeon in People's China (1969) 710

JAMES, I.M.
Visit to China (1967) 259

THE JAPAN ECONOMIC JOURNAL
Peking building huge petro-chemical center (1971) 541

JAPANESE SCIENTIFIC DELEGATION (Natural Science) that visited Communist China
Reports (1966) J 298

KATZMAN, D., H.W. FEELY and Celso BARRIENTOS 322

KESWICK, John
The British Exhibition in Peking, Nov 1964 (1965) 663

KLOCHKO, Mikhail Antonovich
Soviet scientist in China (1964) 373

KRAAR, Louis
I have seen China--and they work (1972) 85

KUO, Zing Yang
Reconstruction in China (1942) 793

KURODA, P.K. and Myint THEIN 356

KURTI, N.
Notes on a visit to China (1965) 263

KWEI, Chi-Ting
The status of physics in China (1944) 794

G.B.L.
C.N. Yang discusses physics in People's Republic of China (1971) 264

LIPPERT, J. and A. AARKROG 307

LÜBECK, Lennart
Electronics in today's China: some impressions from the visit of the Minister of Industry (1972) Sw 510

LUBKIN, Gloria B.
Physics in China (1972) 265

LURCAT, Francois
The People's universities (1972) F 431

LUUKKAINEN, Tapani and Anibal FAUNDES 738

MACKENZIE, Walter C., R.K.C. THOMSON and A.F.W. Peart 729

MADDIN, Stuart
Medicine in China today (1972) 717

MAEGRAITH, Brian
The Chinese are 'liquidating' their disease problems (1957) 718

MARTIN, David
China today (1972) 273

MCFARLANE, Bruce
Letter to the editor (1968) 93

NATURE
Message from Chinese men of science (1943) 805

NEEDHAM, Dorothy and Joseph 819

NEEDHAM, J.
Chinese science (1945) 811

NEEDHAM, J.
Chungking Industrial and Mining Exhibition (1944) 812

NEEDHAM, J.
Science and technology in China's far South-East (1946) 813

NEEDHAM, J.
Science and technology in the North-West of China (1944) 814

NEEDHAM, J.
Science in Chungking (1943) 815

NEEDHAM, J.
Science in Kweichow and Kuangsi (1945) 816

NEEDHAM, J.
Science in South-West China (1943) 817

NEEDHAM, J.
Science in Western Szechuan (1943) 818

NEEDHAM, Joseph and Dorothy
Science outpost (1948) 819

NEEDHAM, J. and HO, Ping-yu 852

NEEDHAM, J. T'ien-ch'in TS'AO and Ping-yu HO 940

OLDHAM, C.H.Geoffrey
The Chinese Trade Exhibition in Japan (1964) 667

OLDHAM, C.H.G.
Earth sciences in the People's Republic of China (1967) 293

OLDHAM, C.H.G.
Science and superstition (1965) 458

OLDHAM, C.H.G.
Science in Mainland China: a tourist's impressions (1965) 181

OLDHAM, C.H.G.
Visits to Chinese institutes of earth science (1965) 294

OLDHAM, C.H.G.
Visits to Chinese research institutes and scientific instrument factories (1965) 305

OLDHAM, C.H.G.
Visits to Chinese schools
(1965) 459

OLDHAM, C.H.G.
Visits to Chinese universities
(1965) 437

OLIPHANT, Mark
Over pots of tea: excerpts
from a diary of a visit to
China (1966) 267

PEART, A.F.W., R.K.C. THOMSON
and Walter C. MACKENZIE 729

PENFIELD, Wilder
Oriental renaissance in
education and medicine (1963) 721

PERSSON, G.
Fractionation phenomena in
debris from the Chinese nuclear
explosion in May 1965 (1969) 348

PERSSON, G.
Observations on debris from the
first Chinese nuclear test
(1969) 349

PERSSON, G. and J. SISEFSKY
Debris from the sixth Chinese
nuclear test (1969) 350

RICHARDSON, S. Dennis
A modern-day Marco Polo visits
China (1965) 492

RIVIERE, BLAIN and ALLARD 599

SCIENCE AND TECHNOLOGY IN CHINA
National Academy of Peiping
(1948) 821

SEWELL, W.G.
A chemist in China (1972) 822

SHERRARD, Howard
Notes on road and bridge
engineering in China (1968) 636

SIDEL, Victor W.
Serve the people: medical care
in the People's Republic of
China (1972) 725

SIGNER, Ethan
New directions in Chinese
science (1971) 189

SIGNER, E. and Arthur W. GALSTON
Education and science in China
(1972) 190

SIGURDSON, Jon
Rural industry--a traveller's
view (1972) 107

SILOW, R.A.
The Scientific activities of
the British Council in China
(1948) 824

SISEFSKY, J.
Debris particles from the
second Chinese nuclear bomb
test (1966) 355

SISEFSKY, J. and G. PERSSON 350

SIVIN, Nathan
Chinese alchemy: preliminary
studies (1968) 934

STUDY GROUP FOR SCIENCE AND
TECHNOLOGY OF NEIGHBOURING
NATIONS, Foreign Machine
Industry Survey Committee
Analysis of present status of
Communist China's science and
technology (1965) 590

SUNG, Ying-hsing
T'ien-kung K'ai-wu: Chinese
technology in the seventeenth
century (1966) 939

TANG, Pei-sung
Biology in war-time China
(1944) 827

THEIN, Myint and P.K. KURODA
Global circulation of radiocerium isotopes from the May 14
1965 nuclear explosion (1967) 356

THOMPSON, H.W.
Science in China (1963) 198

THOMSON, R.K.C., Walter C. MACKENZIE and A.F.W. PEART
A visit to the People's Republic of China (1967) 729

TING, V.K.
Scientific research in China (1935) 829

TS'AO, T'ien-ch'in, HO, Ping-yu and Joseph NEEDHAM
An Early Mediaeval Chinese alchemical text on aqueous solutions (1959) 940

TSU, Raphael
High technology in China (1972) 516

US NEWS AND WORLD REPORT
A look at science in Red China (1958) 255

VON HOFSTEN, Bengt
Chemistry and politics in the People's Republic of China (1972) Sw 288

WADDINGTON, C.H.
Biology in China (1963) 284

WASHINGTON SCIENCE TRENDS
Mainland China computers (1972) 517

WEDGWOOD BENN, Anthony
China--land of struggle. Criticism and transformation (1972) 205

WILLOX, G.L.
Contemporary Chinese health, medical practice and philosophy (1966) 731

WILSON, Dick
Apricot-time in Anshan (1964) 611

WILSON, D.
Black crow into peacock (1964) 574

WILSON, D.
Chemicals for the communes (1964) 529

WILSON, D.
Fuel from Fushun (1964) 563

WILSON, D.
No more by line of skin (1964) 591

WILSON, D.
Plastics in Shanghai (1964) 530

WILSON, D.
Presses from Shenyang (1964) 592

WILSON, D.
Shanghai machine tools (1964) 593

WILSON, J. Tuzo
Geophysical institutes of the USSR and of the People's Republic of China (1959) 300

WILSON, J.T.
Geophysics (1961) 301

WILSON, J.T.
Mao's almanac (1972) 302

WILSON, J.T.
One Chinese moon (1959) 208

WILSON, J.T.
Red China's hidden capital of science (1958) 303

YANG, Chen Ning
Education and scientific research in China (1972) 212

YONG, Lam-lay
The geometrical basis of the ancient Chinese square-root method (1970) 944

ABE, Munemitsu
Spare-time education in Communist China (1961) 400

ABELSON, P.H.
Mainland China: an emerging power (1967) 401

ABELSON, Phillip H.
The Chinese A-Bomb (1964) 308

ADLER, Solomon
The Chinese economy (1957) 1

AGREN, Hans
Medical practice in China: a compendium (1972) 683

AIRD, John S.
Population policy and demographic prospects in the People's Republic of China (1972) 734

AIRD, John and Lawrence KRADER 678

AKABANE, Nebuhisa
Chemical industry: technological self-reliance preceded the development stage (1972) J 519

AKABANE, N. and Reiitsu KOJIMA
China's chemical industry (1966) J 520

AKAGI, Akio
Large-scale electronic computers for general use (1972) J 504

AKENO, Yoshio
The Chinese iron and steel industry (1964) J 598

AKENO, Y.
The Chinese transportation industry (1966) J 638

ALLEN, Edwin J. Jr.
Disease control in China (1965) 684

ANDORS, Stephen
Revolution and modernization: man and machine in industrializing society, the Chinese case (1969) 59

ARTS AND SCIENCES IN CHINA
A science-minded nation (1963) 403

ARTS AND SCIENCES IN CHINA
Science notes (quarterly 1963/64) 115

ASHTON, John
Development of electric energy resources in Communist China (1968) 496

AU, Lewis Li-tang
Civil and hydraulic engineering (1961) 631

AUDETTE, Donald G.
Computer technology in Communist China 1956-1965 (1966) 505

AUTOMOTIVE INDUSTRIES
China's liberation model truck goes into production (1956) 595

AVIATION WEEK & SPACE TECHNOLOGY
Chinese push ICBM development (1967) 309

BARNETT, A. Doak
The inclusion of Communist China in an arms-control program (1960) 310

BARNETT, A.D.
Mao versus modernization (1968) 214

BARNETT, James W. Jr.
What price China's bomb? (1967) 116

BASALLA, George
The spread of Western science (1967) 767

BASTID, Marianne
Economic necessity and political ideals in educational reform during the Cultural Revolution (1970) 404

BAUM, Richard D.
'Red and Expert': the politico-ideological foundations of China's Great Leap Forward (1964) 215

BEATON, Leonard
The Chinese and nuclear weapons (1962) 311

BEATON, L.
The Chinese bomb (1965) 312

BEATON, L. and John MADDOX
The spread of nuclear weapons (1962) 313

BEER, A., HO Ping-yu, LU Gwei-Djen, J. NEEDHAM, E.G. PULLEYBLANK, G.I. THOMPSON
An 8th-century Meridian Line: I-Hsing's chain of gnomons and the pre-history of the metric system (1961) 834

BEER, A., Joseph NEEDHAM and HO Ping-yu 916

BENNETT, Adrian Arthur
John Fryer: the introduction of Western science and technology into nineteenth century China (1967) 768

BERBERET, John A.
The prospects for Chinese science and technology (1968) 117

BERBERET, J.A.
Science and technology in China (1972) 118

BERBERET, J.A.
Science and technology in Communist China (1960) 119

BERBERET, J.A.
Science, technology and Peking's planning problems (1962) 120

BERGEMAN, Thomas H.
The origins of modern science in China: research in chemistry and mathematics before 1938 (n.d.) 769

BERGER, Roland
Self-reliance, past and present (1970) 60

BERI, G.C.
The development of education and education and professional manpower in China and India: a comparative study (1969) 405

BERNAL, J.D.
Science and technology in China (1956) 406

BERNAL, J.D.
Science in history (1965) 835

BERNARD, Henri
Matteo Ricci's scientific contribution to China (1935) 770

BEYER, Robert T.
Solid state physics (1961) 261

BIGGERSTAFF, Knight
The earliest modern government schools in China (1961) 771

BIGGERSTAFF, K.
Shanghai Polytechnic Institution and Reading Room: an attempt to introduce Western science and technology to the Chinese (1956) 772

BOORMAN, Howard L.
The Scientific Revolution in Communist China (1968) 122

BOSWELL, P.G.H.
Geology in China (1942) 773

BRENNAN, Charles
China: 'King' cotton (1966) 570

BRULE, Jean-Pierre
China comes of age (1971) F 123

BUCK, J.
Fact and theory about China's land (1949) 464

BUSINESS WEEK
China's push to catch up in science (1966) 124

BUSINESS WEEK
China's shot tells a startling story (1964) 314

BUSINESS WEEK
Red China's beehive of science (1959) 125

BUSINESS WEEK
Red China's 'leap' toward science (1962) 126

CALDER, Nigel
Mao--1 in orbit (1970) 315

CARIN, Robert
Irrigation scheme in Communist China (1963) 751

CARIN, R.
Power industry in Communist China (1969) 497

CARIN, R.
River control in Communist China (1962) 752

CARTIER, Michel
Planning of education and professional training in Mainland China (1965) F 407

CASELLA, Alessandro
China's atomic tiger (1966) 316

CHANG, Alfred Zee
Scientists in Communist China (1954) 216

CHANG, K.S.
Geographical bases for industrial development in Northwestern China (1962) 648

CHANG, Parris H.
China's scientists in the Cultural Revolution (1969) 217

CHAO, E.C.T.
Progress and outlook of geology (1961) 289

CHAO, Kang
Agricultural production in Communist China, 1949-1965 (1970) 465

CHAO, K.
The construction industry in Communist China (1968) 575

CHAO, K.
The rate and pattern of industrial growth in Communist China (1965) 61

CHAO, Kuo-chun
Rural manpower in India and China (1962) 671

CHEMICAL & ENGINEERING NEWS
China: few scientific contacts (1971) 365

CHEMICAL & ENGINEERING NEWS
China: science on a swinging pendulum (1972) 218

CHEMICAL & ENGINEERING NEWS
Chinese scientists in midst of US tour (1972) 366

CHEMICAL ENGINEERING PROGRESS
Nuclear progress in Red China (1960) 127

CH'EN Ch'i-t'ien (Gideon)
Pioneer promoters of modern industrial technique in China (1934-1938) 775

CH'EN Feng-chi
Scientific and technical education in China (1963) J 408

CH'EN, Gideon, pseud.; Ch'en Ch'i-t'ien 775

CHEN, Jack
Taking off on a tricycle (1971) 62

CHEN, Ke-Chung
Petroleum industry in the Chinese Mainland (1964) 533

CHEN, Nai-ruenn and Walter GALENSON
The Chinese economy under Communism (1969) 4

CHEN, Paul
Acupuncture and moxibustion (1966) 689

CHEN, Pi-chao
China's birth control action programme, 1956-1964 (1970) 735

CHEN, Pi-chao
Overurbanization, rustication of urban-educated youths, and politics of rural transformation (1972) 672

CHEN, Theodore H.E.
Science, scientists and politics (1961) 219

CHEN, T.H.E.
Thought reform of the Chinese intellectuals (1960) 220

CHEN, T.H.E.
Education in Communist China (1970) 409

CHEN, William Y.
Medicine and public health (1961) 690

CHENG, Chu-Yuan
Economic relations between Peking and Moscow: 1949-63 (1964) 44

CHENG, C.Y.
The machine building industry in Communist China (1972) 576

CHENG, C.Y.
Peking's minds of tomorrow: problems in developing scientific and technological talent in China (1966) 410

CHENG, C.Y.
Progress of nuclear weapons in Communist China (1965) 317

CHENG, C.Y.
Scientific and engineering manpower in Communist China (1968) 411

CHENG, C.Y.
Scientific and engineering manpower in Communist China 1949-1963 (1965) 412

CHENG, Huan
There is something to it (1972) 691

CHENG, Tien-hsi
Disease control and prevention in China (1972) 692

CHENG, T.H.
The entomological society of (Communist) China (1963) 367

CHENG, T.H.
Insect control in Mainland China (1963) 466

CHENG, T.H.
Production of kelp--a major aspect of China's exploitation of the sea (1969) 467

CHENG, T.H.
Schistosomiasis in Mainland China (1971) 693

CHENG, T.H.
Utilization of wild plants in Communist China (1965) 649

CHENG, T.H.
Zoological sciences since 1949 (1961) 286

CHENG, T.H. and Roy H. DOI
Recent nucleic acid research in China (1968) 277

CHESNEAUX, J.
Science in the Far East (1966) 776

CHESNEAUX, J.
The spread of modern science in the Far East (1966) 128

CHESNEAUX, J. and J. NEEDHAM
Science in the Far East from the 16th to the 18th century (1965) 777

CH'IEN, Cheng
A dragon with nuclear teeth (1966) 129

CHINA NEWS ANALYSIS
Training in medicine (1965) 413

CHINA NEWS ANALYSIS
University professors (1955) 134

CHINA NEWS ANALYSIS
Weights and measures: industrial standards (1962) 462

CHINA REPORT
The Father of Chinese rocketry (1967) 226

CHINA REPORT
Higher education in science and technology in China (1966) 414

CHINA REPORT
Two demands on science (1965) 227

CHINA RESEARCH INSTITUTE (Tokyo)
Medicine and hygiene in New China (1952) J 697

CHINA RESEARCH INSTITUTE
Present conditions of the machine industry in China (1955) J 577

CHINA RESEARCH INSTITUTE
Reform of agricultural technology in China (1954) J 468

CHINA RESEARCH INSTITUTE
Technical development of Chinese heavy industry (1957) J 578

CHINA RESEARCH INSTITUTE
The 1964 Peking Science Symposium 1 (1964) J 369

CHRISTIANSEN, W.N.
Science and the scientist in China today (1968) 135

CIENCIA Y TECNICA EN EL MUNDO
Scientific and industrial development in Mainland China (1971) Sp 64

CLOSE, Alexandra
China streamlines her steel (1965) 600

CLOSE, A.
Down to earth (1966) 522

COMBRIDGE, J.H.
The Chinese water-balance escapement (1964) 837

CROIZIER, Ralph C.
Chinese Communist attitudes toward traditional medicine (1966) 698

CROIZIER, R.C.
Traditional medicine in Communist China; science, Communism and cultural nationalism (1965) 699

CROIZIER, R.C.
Traditional medicine in Modern China: science, nationalism, and the tensions of cultural change (1968) 700

CURRENT SCENE
China's 'reformed' universities: the first year (1971) 415

CURRENT SCENE
Chinese science on the mend (1971) 136

CURRENT SCENE
The conflict between Mao Tse-tung and Liu Shao-ch'i over agricultural mechanization in Communist China (1968) 32

CURRENT SCENE
Decision for an 'upsurge': some impressions on Peking's approach to economic problems (1965) 47

CURRENT SCENE
Industrial development in China: a return to decentralization (1968) 65

CURRENT SCENE
Mao's revolution in public health (1968) 701

DAGENAIS, F.
Science in Early Republican China: the development of scientific societies, 1914-1927 (1964) 779

DAVIS, Tenney L. and CH'EN Kuo-fu Kuo-fu
Shang yang-tzu, Taoist writer and commentator on alchemy (1942) 838

DAWSON, Owen L.
China's two-pronged agricultural dilemma: more chemical fertilizer and irrigation needed for more food (1965) 469

DAWSON, O.L.
Communist China's agriculture, its development and future potential (1970) 470

DAWSON, O.L.
Irrigation developments under the Communist Regime (1966) 753

DEAN, Genevieve C.
A note on the sources of technological innovation in the People's Republic of China (1972) 633

DEAN, G.C.
Innovation in the choice of techniques context: the Chinese experience, 1958-1970 (1972) 66

DEAN, G.C.
Science and the thought of Chairman Mao (1970) 137

D'ELIA, Pasquale M.
Galileo in China (1960) 780

DERNBERGER, Robert F.
The relationship between foreign trade, innovation, and economic growth in Communist China (1968) 48

DERNBERGER, R.F.
Economic realities (1966) 33

DERNBERGER, R.F.
Economic realities and China's political economics (1969) 34

DELEYNE, Jan
The Chinese economy (1971) F 138

DIETRICH, Craig
Cotton culture and manufacture in Early Ch'ing China (1972) 839

DIMOND, E. Grey
Medical education in China (1972) 417

DOI, Roy H. and Tien-hsi CHENG 277

DONNITHORNE, Audrey
China's economic system (1967) 7

DOOLIN, Dennis J.
'Both Red and Expert': the dilemma of the Chinese intellectual (1963) 232

DOUGLAS, Robert K.
The progress of science in China (1873) 781

DUCROCQ, Albert
The Chinese satellite (1970) F 318

DUNCAN, James S.
Red China's economic development since 1949 (1966) 8

DURDIN, Peggy
Medicine in China: a revealing story (1960) 703

DWYER, D.J.
The coal industry in Mainland China since 1949 (1963) 613

EBERHARD, Wolfram
The political function of astronomy and astronomers in Han China (1957) 840

ECKSTEIN, Alexander
Strategy of economic development in Communist China (1961) 35

ECKSTEIN, A.
Communist China's economic growth and foreign trade: implications for US policy (1966) 36

ECKSTEIN, A.
Manpower and industrialization in Communist China, 1952-1957 (1960) 675

ECKSTEIN, A., John K. FAIRBANK and L.S. YANG 783

ECONOMIC RESEARCH INSTITUTE
Analysis of present state of industrial technology in China (2) (1968) J 579

THE ECONOMIST
Mao calls the boffins home (1956) 233

EHRLICH, Paul R. and John P. HOLDREN
Neither Marx nor Malthus (1971) 737

EITNER, Hans-Jurgen
Education and science in the Chinese People's Republic, 1949-1963 (1964) G 139

ELECTRONIC DESIGN
Red Chinese are turning out computers with 'modest' ICs (1972) 506

ELECTRONICS INTERNATIONAL
China poised for 'Great Leap' into the forefront of science (1969) 507

ELLIOT, Denis
Spare-time studies (1964) 418

ELVIN, Mark
The high-level equilibrium trap: the causes of the decline of invention in the traditional Chinese textile industries (1972) 841

ELVIN, M.
The State, printing and the spread of scientific and technical knowledge in China, 950-1350 (1971) 842

EMERSON, John Philip
Employment in Mainland China: problems and prospects (1968) 419

EMERSON, J.P.
Manpower absorption in the non-agricultural branches of the economy of Communist China, 1953-1958 (1961) 676

EPINAT'EVA, A.M. and I.P. KOSMINSKAYA
Seismic prospecting in China (1959) 364

ERISMAN, Alva Lewis
China: agricultural development, 1949-1971 (1972) 471

ERSELCUK, Muzzaffer
The iron and steel industry in China (1956) 601

ESPOSITO, Bruce J.
The Cultural Revolution and science policy and development in Mainland China (1971) 234

ESPOSITO, B.J.
The People's Liberation Army, medicine and the Cultural Revolution (1971) 705

ESPOSITO, B.J.
The politics of medicine in the People's Republic of China (1972) 706

ESPOSITO, B.J.
Science in Mainland China (1972) 235

ESSO STANDARD S.A.F. Departement Information
The petroleum industry in People's China (1968) F 535

EVANS, Gordon Heyd
China and the atom bomb (1962) 320

EVANS, G.H.
Communist China's A-Bomb program (1961) 321

FAIRBANK, John K.
The influence of modern Western science and technology on Japan and China (1955) 782

FAIRBANK, J.K., Alexander ECKSTEIN and L.S. YANG
Economic change in Early Modern China: an analytic framework (1960) 783

FAR EAST TRADE & DEVELOPMENT
China-UK understanding (1972) 372

FAR EAST TRADE & DEVELOPMENT
How China is managing (1972) 67

FAR EAST TRADE & DEVELOPMENT
Modernizing coal production (1969) 614

FAR EAST TRADE & DEVELOPMENT
Towards positive trade relations (1972) 660

FAR EASTERN ECONOMIC REVIEW
Deals without dependence (1971) 661

FAR EASTERN ECONOMIC REVIEW
A new priority for steel (1971) 602

FAR EASTERN ECONOMIC REVIEW
Technical license for China (1962) 571

FAR EASTERN ECONOMIC REVIEW
Tokyo-Peking trade winds (1965) 662

FENG, Wen
Peiping's financial burden in developing nuclear weapons (1967) 323

FISHANE, S.J.
China's technical position (1960) 141

FIX, Joseph E. III
China: the nuclear threat (1966) 324

FOO-KUNE, C.F.
Science and industry in China (1960) 142

FRANK, Lewis A.
Nuclear weapons development in China (1966) 325

FREEBERNE, Michael
Birth control in China (1964) 739

FREEBERNE, M.
The spectre of Malthus: birth control in Communist China (1963) 740

FUJII, Shoji
China's nuclear scientists (1964) 326

FUNG, Yu-lan
Why China has no science--an interpretation of the history and consequences of Chinese philosophy (1922) 844

FURTH, Charlotte
Ting Wen-chiang: science and China's new culture (1970) 784

GALENSON, Walter and Nai-ruenn CHEN 4

GALLAGHER, Louis J.
China in the 16th century: the journals of Matthew Ricci (1942) 845

GARDNER, Frank J.
Chinese oil flow up, but much larger gains needed (1971) 536

GARRATT, Colin
China as a foreign aid donor (1961) 759

GAUVENET, Andre
The Chinese nuclear tests (1967) F 327

GEALY, Edgar J. and Anton W.T. WEI
Mainland China (1965, 1966, 1967) 615

GOLDEN, Ronald
Peking pushes for an expanded international air transport system (1971) 639

GOODSTADT, Leo F.
China mounts a war on pests (1972) 472

GOODSTADT, L.F.
From the land, a new power struggle (1972) 508

GOODSTADT, L.F.
Leaping backwards (1969) 49

GRAY, Jack
The economics of Maoism (1969) 37

GUILLAIN, Robert
Ten years of secrecy (1965) 328

HALPAP, P.
The mining industry in the Chinese People's Republic (1963) 616

HALPERIN, Morton H.
China and the bomb (1965) 329

HALPERIN. M.H.
China's nuclear strategy (1966) 330

HAN, Suyin
Acupuncture--the scientific evidence (1964) 708

HAN, S.
China in the year 2001 (1970) 69

HAN, S.
Family planning in China (1970) 741

HAN, S.
Family planning in China today (1965) 742

HAO, Paul L.C.
The Chinese Communist petroleum industry (1967) 537

HARARI, Roland
The long march of Chinese science (1968) 145

HARASHINA, Kyoichi
Growth centered around tractor building (1972) J 580

HARLAND, W.B.
The organization of geology overseas: China (1966) 291

HARPER, Paul
Closing the education gap: problems of industrial spare-time schools (1965) 423

HARPER, P.
Spare-time education for workers in Communist China (1964) 424

HASKINS, Caryl P.
The Scientific Revolution and world politics (1964) 146

HAUDRICOURT, A. and Joseph NEEDHAM
Ancient Chinese science (1965) 846

HEENAN, Brian
China's petroleum industry (1965) 538

HEENAN, L.D.B.
The Chinese petroleum industry (1966) 539

HEENAN, L.D.B.
The petroleum industry of Monsoon Asia (1965) 540

HEMY, G.W.
The Chinese coal industry (1961) 617

HERON, Antoine
Revolutionary management of enterprises (1971) F 70

HIEDA, Kentaro
Concerning medical science in China (1970) J 709

HINIKER, Paul J. and R. Vincent FARACE
Approaches to national development in China: 1949-1958 (1969) 38

HINTON, William
Iron oxen: a documentary of revolution in Chinese farming (1970) 473

HIRAMATSU, Shigeo
China's Socialist construction and 'self-reliance' (1972) J 71

HO, Peng-yoke
Ancient Chinese astronomical records and their modern applications (1970) 847

HO, P.Y.
The birth of modern science in China (1967) 848

HO, P.Y. and Joseph NEEDHAM
Ancient Chinese observations of solar haloes and parhelia (1959) 849

HO, P.Y. and Joseph NEEDHAM
Elixir poisoning in Mediaeval China (1959) 850

HO, P.Y. and Joseph NEEDHAM
The laboratory equipment of the Early Mediaeval Chinese alchemists (1959) 851

HO, P.Y.and Joseph NEEDHAM
Theories of categories in Early Mediaeval Chinese alchemy (1959) 852

HO, P.Y., Joseph NEEDHAM and Arthur BEER 916

HO, P.Y. TS'AO T'ien-ch'in and Joseph NEEDHAM 940

HODGES, Henry
Technology in the Ancient World (1970) 853

HOFFMAN, Charles
Work incentive practices and policies in the People's Republic of China, 1953-65 (1967) 72

HOLDREN, John P. and Paul R. EHRLICH 737

HOPKINS, C.E. and J.E. STEPANEK
China's AIS--a point four pioneer; tested method of increasing output in underdeveloped lands by putting better tools in hands of average peasant (1949) 786

HORN, Joshua S.
Quantity and quality in surgery (1963) 711

HORN, J.S.
Breakthrough tactics in Chinese surgery (1964) 712

HOSHINO, Yoshiro
Basic problems of technological innovation (1969) J 73

HOSHINO, Y.
China's science and technology and Japan's science and technology (1972) J 74

HOSHINO, Y.
China's technological line in the Great Proletarian Cultural Revolution (1971) J 75

HOU, Chi-ming
Economic dualism: the case of China, 1840-1937 (1963) 787

HOU, C.M.
External trade, foreign investment, and domestic development: the Chinese experience, 1840-1937 (1961) 788

HOU, C.M.
Foreign investment and economic development in China, 1840-1937 (1965) 789

HOWE, Christopher
Problems, performance and prospects of the Chinese economy (1967) 76

HSIA, R.
Anshan steel (1965) 603

HSIA, R.
China's key steel bases (1966) 604

HSIA, R.
Paotow steel (1966) 605

HSIA, R.
Wuhan steel (1966) 606

HSIA, Ronald
The development of Mainland China's steel industry since 1958 (1961) 607

HSIAO, Chi-jung
Production and supply of chemical fertilizers on the Chinese Mainland (1965) 524

HSIEH, Chiao-min
Hsia-ke Hsu--Pioneer of modern geography in China (1958) 854

HSIEH, C.M.
The status of geography in Communist China (1959) 292

HSU, Francis L.K.
A cholera epidemic in a Chinese town (1955) 790

HSU, F.L.K.
Religion, science, and human crises: a study of China in transition and its implications for the West (1952) 791

HSU, Immanuel C.Y.
The impact of industrialization on higher education in Communist China (1965) 426

HU, C.T.
Communist education: theory and practice (1962) 427

HU, Chang-tu
Chinese education under Communism (1962) 428

HU, Chang-tu, et al
China: its people, its society, its culture (1960) 713

HU, Shih
The Chinese Renaissance (1934) 855

HU, S.
The scientific spirit and method in Chinese philosophy (1967) 856

HUARD, Pierre and Ming WONG
French enquiry into Chinese science and technology in the 18th century (1966) F 857

HUARD, P. and M. WONG
Evolution of Chinese materia medica (1958) F 858

HUARD, P. and M. WONG
Development of technology in China at the 19th century (1962) F 859

HUARD, P. and M. WONG
China, yesterday and today (1960) F 860

HUARD, P. and M. WONG
Chinese medicine (1968) 714

HUARD, P. and M. WONG
Taoism and science (1956) F 861

HUGHES, E.R-
The invasion of China by the Western world (1937) 792

HUNTER, Holland
Transport in Soviet and Chinese development (1965) 640

IKLE, F.C.
The growth of China's scientific and technical manpower (1957) 429

INDUSTRIAL RESEARCH
Red China science focuses on short-term (1971) 148

INGLIS, David R.
The Chinese bombshell (1965) 331

INSTITUT FÜR ASIENKUNDE (Hamburg)
Economic relations of the PRC with the Soviet Union (1959) G 50

ISHIKAWA, Shigeru
Change of employment and structure of productivity in Communist China (1961) J 677

ISHIKAWA, S.
The Chinese economy: a general framework for long-term projection (1966) 10

ISHIKAWA, S.
Choice of techniques in Mainland China (1962) 28

ISHIKAWA, S.
Long-term outlook for Chinese agriculture (1967) J 474

ISHIKAWA, S.
A note on the choice of technology in China (1972) 29

ISHIKAWA, S.
Recent changes in Chinese agricultural techniques (1962) J 475

JAUBERT, Alain
Research and development in China (1971) 150

JOHNSON, Chalmers
China's 'Manhattan Project' (1964) 332

JOHNSTON, Douglas M. and Hungdah CHIU
Agreements of the People's Republic of China, 1949-1967: a calendar (1968) 760

JONAS, Anne M.
Atomic energy in Soviet Bloc Nations (1959) 333

JONES, P.H.M.
Creeping modernization (1964) 476

JONES, P.H.M.
Machines on the farm (1964) 477

JONES, P.H.M.
One million tractors (1964) 478

KAMBARA, Shu (ed.)
The chemical industry in China (1970) J 525

KAWAMURA, Yoshio
Supply of and demand for lumber and afforestation policy in China (1966) J 479

KAWAMURA, Yoshio
Technological revolution in Chinese agriculture (1965) J 480

KIANG, T.
Possible dates of birth of Pulsars from ancient Chinese records (1969) 863

KIRBY, E. Stuart
Trade and development of Mainland China (1968) 664

KIRK, Don
China's three way stretch (1967) 509

KISHIDA, Junnosuke
Chinese nuclear development (1967) 334

KLATT, Werner, ed.
The Chinese model (1965) 51

KLEIN, Sidney
Sino-Soviet economic relations, 1949-62: a sinologist's sketch (1963) 52

KOBAYASHI, Fumio
Organization of scientific and technical research (1967) J 151

KOBAYASHI, F.
Development of the 'Scientific Experiment' movement in China and the problem of training scientific and technical manpower (1965) J 430

KOJIMA, Reiitsu
The agricultural machinery and implement industry (1967) J 581

KOJIMA, R.
Agricultural machinery and tools industry in the development of a self-sustaining national economy (1966) J 582

KOJIMA, R.
China's indigenous technology (1972) J 78

KOJIMA, R.
The Chinese machine tool industry (1966) J 583

KOJIMA, R.
The direction of technology in China (1968) J 79

KOJIMA, R.
On an 'integrated national democratic economy' (1965) J 80

KOJIMA, R.
On intelligent technology: indigenous thought supports China's economic base (1972) J 81

KOJIMA, R.
The iron and steel industry (1970) J 608

KOJIMA, R.
Mao Tse-tung's thoughts on science and technology (1972) J 82

KOJIMA, R.
Reappraisal of the Great Leap Forward policy: with special reference to the industrialization of rural economy (1967) J 83

KOJIMA, R.
Recovery of nature: construction in mountain areas before the Great Leap Forward (1972) J 754

KOJIMA, R.
'Self-sustained national economy' in Mainland China (1967) 84

KOJIMA, R.
Textile industry (1967) J 572

KOJIMA, Reiitsu and Nobuhisa AKABANE 520

KOJIMA, Reiitsu, Masahisa SUGANUMA and Kazuo YAMANOUCHI 109

KOLB, John and Sol SANDERS 99

KONTAKT
Chinese researchers must serve China--not themselves (1968) Da 236

KOVNER, Milton
Communist China's foreign aid to less developed countries (1968) 761

KRADER, Lawrence and John AIRD
Sources of demographic data on Mainland China (1959) 678

KRAMISH, Arnold
The Chinese People's Republic and the bomb (1960) 335

KRAMISH, A.
The Great Chinese Bomb Puzzle--and a solution (1966) 336

KU, Y.H.
A survey of Chinese culture--Science (1964) 864

KUMASHIRO, Yukio
On reformation of agricultural technology during the Great Leap Forward period (1968) J 481

KUO, Leslie T.C.
Agricultural mechanization in Communist China (1964) 482

KUO, L.T.C.
Industrial aid to agriculture in Communist China (1967) 483

KUO, L.T.C.
The technical transformation of agriculture in Communist China (1972) 484

KUO, L.T.C. and Ralph W. PHILLIPS 306

KUSANO, Fumiko
Evaluation of comprehensive study of Chinese Communist science and technology (1964) J 152

KWOK, D.W.Y.
Scientism in Chinese thought 1900-1950 (1965) 795

LACOUTURE, Jean
Mao Tse-tung's medicine (1972) F 715

LAPP, Ralph E.
China's mushroom cloud casts a long shadow (1968) 337

LARSEN, Marion R.
China's agriculture under Communism (1968) 485

LATTIMORE, Owen
The industrial impact on China, 1800-1950 (1960) 796

LAURENT, Philippe
Ideological principles of industrial policy (1971) F 86

LEAR, John
Dispassionate scientific look at China (1958) 376

LEAR, J.
Global pollution--the Chinese influence (1971) 377

LEE, Renssalaer W. III
The Hsia Feng System: Marxism and modernization (1966) 237

LEE, R.W. III
Ideology and technical innovation in Chinese industry, 1949-1971 (1972) 87

LEE, Tsung-ying
Synthesizing Chinese and Western medicine (1969) 716

LETHBRIDGE, Henry
China: collectivization and mechanization (1963) 486

LETHBRIDGE, H.
Tractors in China (1963) 487

LETHBRIDGE, H.
Trends in Chinese agriculture (1963) 488

LETULLIER, Andre
Petroleum in People's China (1967) F 542

LI, C.C.
Genetics and animal and plant breeding (1961) 282

LI, Choh-ming
China's industrial development, 1958-63 (1964) 12

LI, Hui-lin
Botanical sciences (1961) 285

LI, Hui-min
An analysis of Peiping's two recent nuclear tests (1967) 338

LI, Yao-tzu and Way Dong WOO
Progress in electronics 1949-1959 (1961) 272

LIM, Robert K.S. and G.H. WANG
Physiological sciences (1961) 283

LINDBECK, John M.H.
An isolationist science policy (1969) 153

LINDBECK, J.M.H.
The organization and development of science (1961) 154

LINDBECK, J.M.H.
Chinese science: it's not a paper atom (1967) 155

LING, H.C., Y.-L. WU and Grace Hsia WU 647

LING, H.C. and Yuan-li WU 656

LIPPITT, Victor D.
Development of transportation in Communist China (1966) 641

THE LISTENER
Science in China (1945) 797

LIU, Chi-chuen
A study of the Chinese Communist nuclear program (1964) 339

LIU, Jung-chao
Fertilizer application in Communist China (1965) 489

LU, Gwei-Djen
China's greatest naturalist: a brief biography of Li Shih-Chen (1966) 865

LU, G.D.
The Inner Elixir (Nei Tan): Chinese physiological alchemy (1972) 866

LU, G.D. and Joseph NEEDHAM
China and the origin of examinations in medicine (1963) 867

LU, G.D. and J. NEEDHAM
A contribution to the history of Chinese dietetics (1951) 868

LU, G.D. and J. NEEDHAM
Medieval preparations of urinary steroid hormone (1969) 869

LU, G.D. and J. NEEDHAM
Records of diseases in Ancient China (1967) 870

LU, G.D., Raphael A. SALAMAN and J. NEEDHAM
The wheelwright's art in Ancient China (1959) 871

LU, G.D. and J. NEEDHAM 917

LU, G.D. and J. NEEDHAM 918

LU, G.D. and J. NEEDHAM 919

LU, G.D. and J. NEEDHAM 920

LU, G.D. and J. NEEDHAM 921

LU, G.D. and J. NEEDHAM 922

LU, G.D. and J. NEEDHAM 923

LU, G.D. and J. NEEDHAM 924

MACDOUGALL, Colina
China keeps the oil flowing (1967) 543

MACDOUGALL, C.
The advancement of learning (1962) 379

MACDOUGALL, C.
China's industrial upsurge (1965) 88

MACDOUGALL, C.
Eight plants for Peking (1964) 665

MACDOUGALL, C.
Fertilizer drive (1965) 526

MACDOUGALL, C.
Fertility rites (1964) 527

MACDOUGALL, C.
Learn from Shanghai (1964) 89

MACDOUGALL, C.
The Reds and the Experts (1964) 238

MACDOUGALL, C.
The struggle to come (1970) 90

MACFARQUHAR, Roderick
The hundred flowers campaign and the Chinese intellectuals (1960) 239

MACIOTI, Manfredo
Recent developments in science and technology in Communist China (1970) F 156

MACIOTI, M.
Science and technology in Communist China (1969) F 157

MACIOTI, M.
Hands of the Chinese (1971) 159

MACIOTI, M.
Mao enters orbit (1970) I 340

MACIOTI, M.
Scientists go barefoot (1971) I 160

MACIOTI, M.
The system of research in China (1972) I 161

MACKAY, Alan
Science in Asia (1965) 162

MADDOX, John
The Chinese A-Bomb--and who next? (1964) 341

MADDOX J. and Leonard BEATON 313

MANN, Felix
Chinese traditional medicine: a practitioner's view (1965) 719

MARTYNOV, V.V.
The coal industry of China (1959) 618

MARU, Rushikesh
Research and development in India and China: a comparative analysis of research statistics and research effort (1969) 163

MATSUMOTO, Hiroichi
Technological innovation in Communist China (1960) 91

MCARTHUR, H. Russell
Technology and technical education in China (1967) 92

MCELHENY, Victor K.
Total synthesis of insulin in Red China (1966) 287

MCFARLANE, Bruce
Letter to the editor (1968) 93

MCFARLANE, B.
Mao's game plan for China's industrial development (1971) 39

MCFARLANE, B. and E.L. WHEELWRIGHT 20

MD MEDICAL NEWSMAGAZINE
China (1971) 720

MENDELSSOHN, Kurt
Aspects of science in China (1963) 432

MENDELSSOHN, K.
China's little leaps (1961) 94

MENDELSSOHN, K.
The impact of technology (1961) 95

MENDELSSOHN, K.
Science and technology in China (1966) 164

MENDELSSOHN, K.
Science in China (1967) 165

MENDELSSOHN, K.
Science in China (1960) 433

MEYERHOFF, A.A.
Developments in Mainland China, 1949-1968 (1970) 544

MIKHAILOV, I.G.
Ultrasonics in the Chinese People's Republic (1960) 266

MIYASITA, Saburo
A link in the Westward transmission of Chinese anatomy in the later Middle Ages (1967) 872

MODELSKI, J.A.
Communist China's challenge in technology (1958) 166

MOORSTEEN, Richard
Economic prospects for Communist China (1959) 13

MUNTHE-KAAS, Harald
China's mechanical heart (1965) 584

MURPHEY, Rhoads
The non-development of science in traditional China (1947) 873

MURPHEY, R.
The Treaty ports and China's modernization: what went wrong? (1970) 799

MUZAKI, Shotaro
Analysis of the present state of China's electric power industry (1959) J 499

MUZAKI, S.
Development of the Chinese machine industry in the past ten years (1960) J 585

MUZAKI, S.
The technical level of the Chinese machine tools industry (1964) J 586

NAKAOKA, Tetsuro and Keiji YAMADA 453

NAKAYAMA, Shigeru
Characteristics of Chinese astrology (1966) 874

NASH, Ralph C. and Tien-hsi CHENG
Research and development of food resources in Communist China (1965) 304

NATURE
Biological research in China (1942) 800

NATURE
Chemistry in China (1945) 801

NATURE
Does China exist (1967) 384

NATURE
East and West in science (1952) 876

NATURE
Geographical research in China (1944) 802

NATURE
Mathematics in China (1944) 803

NATURE
Medical progress in China (1942) 804

NATURE
Old world: welcome visitors
New world: visitors from China
New world: more cracks in the ice (1972) 385

NATURE
Parasitology in Free China (1943) 806

NATURE
Science and engineering in China (1942) 807

NATURE
Scientific development in China (1947) 808

NATURE
Seismology in China (1944) 809

NEEDHAM, Joseph
Aeronautics in Ancient China (1961) 878

NEEDHAM, J.
Astronomy in Classical China (1962) 879

NEEDHAM, J.
Central Asia and the history of science and technology (1949) 880

NEEDHAM, J.
China, Europe, and the seas between (1970) 881

NEEDHAM, J.
China and the invention of the pound-lock (1963/4) 882

NEEDHAM, J.
China's philosophical and scientific traditions (1963) 883

NEEDHAM, J.
Chinese astronomy and the Jesuit Mission (1958) 810

NEEDHAM, J.
The Chinese contribution to the development of the mariner's compass (1970) 884

NEEDHAM, J.
Understanding past is key to future (1965) 912

NEEDHAM, J.
The unity of science: Asia's indispensable contribution (1970) 913

NEEDHAM, J.
Wheels and gear-wheels in Ancient China (1959) 914

NEEDHAM, J.
Within the four seas: the dialogue of East and West (1969) 915

NEEDHAM, J. Arthur BEER and HO Ping-Yü
'Spiked' comets in Ancient China (1957) 916

NEEDHAM, J. and LU, Gwei-Djen
The earliest snow crystal observations (1961) 917

NEEDHAM, J. and LU, G.D.
Efficient equine harness: the Chinese inventions (1960/65) 918

NEEDHAM, J. and LU, G.D.
The Esculentist Movement in Mediaeval Chinese botany; studies on wild (emergency) food plants (1968) 919

NEEDHAM, J. and LU, G.D.
Hygiene and preventive medicine in Ancient China (1959) 920

NEEDHAM, J. with LU, G.D.
Medicine and Chinese culture (1970) 921

NEEDHAM, J. and LU, G.D.
The optick artists of Chiangsu (1966) 922

NEEDHAM, J. with LU, G.D.
Proto-endocrinology in Medieval China (1966) 923

NEEDHAM, J. and LU, G.D.
Sex hormones in the Middle Ages (1968) 924

NEEDHAM, J. and Kenneth ROBINSON
Waves and particles in Chinese scientific thought (1960) F 925

NEEDHAM, J. with WANG Ling
Science and civilization in China. Vol. 1--Introductory orientations (1954) 926

NEEDHAM, J. with WANG Ling
Science and civilization in China. Vol. 2--History of scientific thought (1956) 927

NEEDHAM, J. with WANG Ling
Science and civilization in China. Vol. 3--Mathematics and the sciences of the Heavens and the Earth (1959); Vol. 4--Physics and physical technology: Pt. I: Physics (1962); Pt. II: Mechanical engineering (1965); Pt. III: Civil engineering and nautics (1971); Vol. 5--Chemistry and industrial chemistry (projected); Vol. 6--Biology, agriculture and medicine (projected) 928

NEEDHAM, J. WANG Ling and Derek J. PRICE
Chinese astronomical clockwork (1956) 929

NEEDHAM, J. WANG Ling and Derek J. de Solla PRICE
Heavenly clockwork: the great astronomical clocks of Medieval China (1960) 930

NEEDHAM, J. and Keiji YAMADA
Science in China (1971) 240

NEEDHAM, J. and J. CHESNEAUX 777

NEEDHAM, J. and HO Ping-yu 849

NEEDHAM, J. and HO Ping-yu 850

NEEDHAM, J, and HO Ping-yu 851

NEEDHAM, J. and HO Ping-yu 852

NEEDHAM, J. and LU Gwei-djen 867

NEEDHAM, J. and LU Gwei-djen 868

NEEDHAM, J. and LU Gwei-djen 869

NEEDHAM, J. and LU Gwei-djen 870

NEEDHAM, J., LU Gwei-djen and
Raphael A. SALAMAN 871

NEEDHAM, J, TS'AO T'ien-ch'in
and HO Ping-yu 940

NETRUSOV, A.A.
Development of the chemical
industry in the Chinese People's
Republic (1959) 528

NEW SCIENTIST
A Chinese bomb that went phut
(1968) 342

NIEH, E.K.
Mechanical engineering (1961) 634

NIIJIMA, Junryo
China's Cultural Revolution and
technical reforms (1958) 171

NIU, Sien-chong
China's mineral wealth (1967) 651

NIU, S.C.
Communist China and its nuclear
weapons (1970) 343

NIU, S.C.
Red China's nuclear might
(1970) 344

NOTES ET ETUDES DOCUMENTAIRES
Bilateral aid from the PRC to
countries on the way to development (1965) F 762

NOTES ET ETUDES DOCUMENTAIRES
Education in the People's
Republic of China (1965) F 435

NOTES ET ETUDES DOCUMENTAIRES
The formation of scientific
and technical cadres in the PRC
(1949-1963) (1969) F 436

NOTES ET ETUDES DOCUMENTAIRES
Organization and development of
science in the PRC (1966) F 172

NOTES ET ETUDES DOCUMENTAIRES
Sources of energy in the PRC
(1961) F 652

LA NOUVELLE CHINE
The struggle against pollution
(1972) F 755

NOVICK, Sheldon
The Chinese bomb (1965) 345

NUCLEONICS
Red China claims to be firmly
in atomic age (1959) 346

NUNN, G. Raymond
Publishing in Mainland China
(1966) 386

NYBERG, P. Russell
Computer technology in
Communist China (1968) 512

THE OIL AND GAS JOURNAL
China blames Russia for oil
troubles (1964) 546

THE OIL AND GAS JOURNAL
China nurses flow back up to
'66 level (1970) 547

THE OIL AND GAS JOURNAL
Chinese oil hurt by 'Revolution'
(1968) 548

THE OIL AND GAS JOURNAL
Chinese Reds claim catalytic
reformer built from scratch
(1967) 549

THE OIL AND GAS JOURNAL
To develop an oilfield, get rid
of ghosts and ogres (1966) 550

THE OIL AND GAS JOURNAL
Red China reports higher production (1969) 553

THE OIL AND GAS JOURNAL
Red China reports Taching output up (1970) 554

THE OIL AND GAS JOURNAL
Taching oil hikes Red Chinese output (1969) 555

OLDHAM, C.H. Geoffrey
The challenge of China (1967) 666

OLDHAM, C.H.G.
China in retrospect (1965) 242

OLDHAM, C.H.G.
China today: science (1968) 174

OLDHAM, C.H.G.
A Chinese conference on tectonics (1962) 299

OLDHAM, C.H.G.
The Peking Science Symposium (1964/65) 387

OLDHAM, C.H.G.
Science and education in China (1966) 176

OLDHAM, C.H.G.
Science and technology in China's future (1968) 177

OLDHAM, C.H.G.
Science for the masses? (1968) 178

OLDHAM, C.H.G.
Science in China (1970) 179

OLDHAM, C.H.G.
Science in China's development (1968) 180

OLDHAM, C.H.G.
Science travels the Mao Road (1968) 182

OLDHAM, C.H.G.
Stirrings of British interest in Chinese science (1961) 388

OLDHAM, C.H.G.
Chinese science and the Cultural Revolution (1968) 176

ONOE, Etsuzo
Choice of industrial location in China (1969) J 642

ONOE, E.
Energy policy: the pattern of development in the 1970s (1972) J 500

ONOE, E.
Regional distribution of industries in China: plan and idea (1965) J 643

ONOYE, Etsuzo
Production of agricultural machinery in China (1962) J 587

ONOYE, E.
The Chinese electric power industry (1964) J 501

ONOYE, E.
The Chinese iron and steel industry (1962) J 609

ONOYE, E.
The Chinese power industry (1966) J 502

ONOYE, E.
Research on production location in China (1971) J 644

ORCHARD, J.E.
Industrialization in Japan, China Mainland, and India--some world implications (1960) 14

ORLEANS, Leo A.
Birth control: reversal or postponement? (1960) 743

ORLEANS, L.A.
China's population: reflections and speculations (1966) 744

ORLEANS, L.A.
China's science and technology: continuity and innovation (1972) 184

ORLEANS, L.A.
Communist China's education--policies, problems, and prospects (1968) 438

ORLEANS, L.A.
Education and scientific manpower (1961) 439

ORLEANS, L.A.
Evidence from Chinese medical journals on current population policy (1969) 745

ORLEANS, L.A.
Medical education and manpower in Communist China (1969) 440

ORLEANS, L.A.
A new birth control campaign? (1962) 746

ORLEANS, L.A.
Population redistribution in Communist China (1959) 679

ORLEANS, L.A.
Manpower absorption in Rural China (1961) 680

ORLEANS, L.A.
Professional manpower and education in Communist China (1961) 441

ORLEANS, L.A.
The recent growth of China's urban population (1959) 681

ORLEANS, L.A.
Research and development in Communist China: mood, management and measurement (1967) 185

ORLEANS, L.A.
What is new in birth control in China (n.d.) 747

ORLEANS, L.A. and Richard P. SUTTMEIER
The Mao ethic and environmental quality (1970) 756

OVDIYENKO, I. Kh.
The new geography of industry of China (1959) 40

PAI, Chen; in Union Research Institute 201

PAN, L.C.
Chemical engineering (1961) 635

PAO, Chin-an
Peiping's capacity for nuclear weaponry (1968) 347

PARKER, David
Travelling Chinese (1971) 53

PEAKE, Cyrus H.
Some aspects of the introduction of modern science into China (1934) 820

PEARSON, G.E.
Minerals in China (1962) 619

PERKINS, Dwight
Agricultural development in China 1368-1968 (1969) 490

PETROLEUM PRESS SERVICE
China's response (1965) 557

PETROLEUM PRESS SERVICE
Openings in China (1972) 558

PETROV, Victor P.
China: emerging World power (1967) 442

PHILLIPS, Ralph W. and Leslie T.C. KUO
Agricultural science (1961) 306

PORCH, Harriett E.
Civil aviation in Communist China since 1949 (1966) 645

PORCH, H.E.
The use of aviation in agriculture and forestry in Communist China (1967) 491

PRICE, Derek J., Joseph NEEDHAM and WANG Ling 929

PRICE, D.J., J. NEEDHAM and WANG Ling 930

PRODUCT ENGINEERING
Red China joins USSR to pioneer new automation techniques (1958) 589

PROGRES SCIENTIFIQUE
Career and mobility of researchers in China (1969) F 243

PROGRES SCIENTIFIQUE
China: scientific and industrial developments in China (1970) F 97

PRICE, R.F.
Education in Communist China (1970) 443

PRYBYLA, Jan S.
Communist China's strategy of economic development 1961-1966 (1966) 244

RANGARAO, B.V.
Science in China (training and utilization) (1966) 444

RAY, Denis
'Red and Expert' and China's Cultural Revolution (1970) 245

REICHERS, Philip D.
The electronics industry of China (1972) 513

RETI, Ladislao
The double-acting principle in East and West (1970) 931

RICHER, Philippe
China and the Third World, 1949-1969 (1971) F 763

RICHMAN, Barry M.
Economic development in China and India: some conditioning factors (1972) 460

RICHMAN, B.M.
Industrial society in Communist China: a first hand study of Chinese economic development and management (1969) 98

RIFKIN, Susan B.
Doctors in the fields (1972) 722

RIFKIN, S.B.
Health services in China (1972) 723

RIFKIN, S.B.
The development and use of nuclear energy in the People's Republic of China (1969) 351

RIGBY, Malcolm
Meteorology, hydrology, and oceanography, 1949-1960 (1961) 299

RISKIN, Carl
Local industry and the choice of techniques in the planning of industrial development in Mainland China (1969) 30

RISKIN, C.
Small industry and the Chinese model of development (1971) 31

ROBINSON, Kenneth and Joseph NEEDHAM 925

RYAN, William L. and Sam SUMMERLIN
The China cloud (1969) 246

D.S.
Year of the dove? (1971) 389

SAITO, Akio
The Great Cultural Revolution and the movement to study philosophy (1971) J 247

SALAFF, Janet
Physician heal thyself (1968) 724

SALAFF, Stephen
A biography of Hua Lo-keng (1972) 248

SALAMAN, Raphael A., LU Gwei-Djen and Joseph NEEDHAM 871

SANDERS, Sol and Joan KOLB
China drives toward her own technology (1959) 99

SATO, Masumi
China's technological development in the last ten years (1967) J 100

SATO, M.
China's technological level and pattern of development (1964) J 101

SATO, M.
Chinese medium and small scale industry seen from the technological angle (1966) J 102

SATO, M.
Course of development of the electronics industry (1970) J 514

SATO, M.
Mining industry, including the history and future of the iron and steel industry (1972) J 620

SATO, M.
The realities and technological trends of China's heavy and chemical industries (1967) 103

SATO, M.
Technological development in China viewed through the electronics industry: an engineer's view (1971) 515

SCHUMAN, Julian
Technology interest (1972) 54

SCIENCE JOURNAL
Chinese nuclear test--her second H-Bomb (1967) 352

SCIENCE JOURNAL
Technology behind China's second bomb (1965) 353

SCIENCE NEWS
China joins the space age (1970) 354

SCIENTIFIC RESEARCH
Two kinds of scientists in China: Maoists do well (1967) 250

SEWELL, R.
A guide to the Chinese steel industry (1960) 610

SEWELL, R.
The problem of Chinese technical literature (1960) 390

SHARP, Ilsa
No ivory towers (1971) 447

SHEEKS, Robert
Science in China today (n.d.) 823

SHEEKS, R. and Yuan-li WU 210

SHEEKS, Robert B.
Science, technology and the Cultural Revolution in China (1967) 252

SHIH, Ch'eng-chih
The status of science and education in Communist China and a comparison with that in USSR (1962) 448

SHIH, Joseph Anderson
Science and technology in China (1972) 188

SHIH, Vincent Y.C.
The state of the intellectuals (1970) 253

SIEH, Marie
Medicine in China: wealth for the State (1964) 726

SIGURDSON, Jon
China: re-cycling that pays (1972) 757

SIGURDSON, J.
Commentary on industrial policy and technological development in China (1972) Sw 105

SIGURDSON, J.
Factories in the fields (1972) 106

SIGURDSON, J.
Natural science and technology in China (1968) Sw 191

SIGURDSON, J.
China's technological development (1970) Sw 41

SIGURDSON, J.
Chinese attitudes toward science (1970) Sw 192

SIVIN, Nathan
Cosmos and computation in early Chinese mathematical astronomy (1969) 933

SIVIN, N
Chinese alchemy: preliminary studies (1968) 934

SLAVEK, P.
Education and research in China (1962) Sw 193

SLEZAK, F.
China: development of the petroleum industry (1970) G 559

SMITH, P.J. and J. NEEDHAM
Magnetic declination in Mediaeval China (1967) 935

SNOW, Edgar
The other side of the river (1961) 194

SNYDER, Charles
Tomorrow's challenge (1970) 758

SOREL, J.J.
China, new scientific power (1966) F 195

SPENCE, Jonathan
The China helpers: Western advisers in China, 1620-1960 (1969) 826

STEPANEK, J.E. and C.E. HOPKINS 786

STEPHENS, Michael Dawson
The coal industry of N.W. and S.W. China (1964) 621

STONE, Marshall H.
Mathematics, 1949-1960 (1961) 260

STOVICKOVA, Dana
What is acupuncture? (1961) 727

STRUBELL, Wolfgang
On extraction and refining of petroleum in Old China (1968) G 936

SUGA, Sakae
The level of China's industrial technology (1967) 108

SUGANUMA, Masahisa, Reiitsu KOJIMA and Kazuo YAMANOUCHI
The direction of development of the Chinese Socialist economy: study of China's local small-scale industry (1971) J 109

SUN, E-tu Zen
Sericulture and silk textile production in Ch'ing China (1972) 937

SUN, E-tu Zen
Wu Ch'i-chun: profile of a Chinese scholar-technologist (1965) 938

SUTTMEIER, Richard P.
Party views of science: the record from the first decade (1970) 254

SUTTMEIER, R.P.
Scientific societies: a chapter in Chinese scientific development (1972) 197

SUTTMEIER, R.P. and Leo A. ORLEANS 756

SWETZ, Frank
Training of mathematics teachers in the People's Republic of China (1970) 449

TAKAHASHI, Keiji YAMADA and Ichii SABURO
Theory and practice as seen in the School of Chinese Medicine (1971) J 728

TAMURA, Saburo
Chemicals and mechanization processed by the mass movement of workers and peasants (1972) J 493

TANSKY, Leo
China's foreign aid: the record (1972) 764

TAWNEY, R.H.
Land and labour in China (1932) 828

TIEN, H. Yuan
Educational expansion, deployment of educated personnel and economic development in China (1967) 682

TIEN, H.Y.
Population control: recent developments in China (1962) 749

TIEN, H.Y.
Sterilization, oral contraception, and population control in China (1965) 750

TRETIAK, Daniel
Has China enough oil? (1964) 560

TSANG, Chiu Sam
Society, schools and progress in China (1968) 450

TSAO, T.C.
Electrical engineering (1961) 637

TSAO, T'ien-ch'in, HO Ping-yu and Joseph NEEDHAM
An Early Mediaeval Chinese alchemical text on aqueous solutions (1959) 940

TSIEN, Tche-hao
Higher education and scientific research in the PRC (1971) F 199

TSUTSUMI, Shigeru
From natural to man-made fibres (1972) J 573

UCHIDA, Genko
Technology in China (1966) 110

UCHIDA, G.
White Paper on present day Chinese technology (n.d.) 111

ULLERICH, Curtis
Size and composition of the Chinese GNP (1972) 17

UNGER, Jonathan
Mao's million amateur technicians (1971) 112

UNGER, J.
On snails and pills (1971) 730

US CONGRESS: Senate
National policy machinery in Communist China (1960) 202

US CONGRESS: Senate
Staffing procedures and problems in Communist China (1963) 451

US NEWS AND WORLD REPORT
A threat to the United States in another 5 or 6 years (1966) 203

L'USINE NOUVELLE
China, petroleum producer . . . the day after tomorrow?(1968) F 561

VETTERLING, Philip W. and James J. WAGY
China: the transportation sector, 1950-71 (1972) 646

VAN RADERS, H.A.
The awakening giant (1965) Da 622

WAKAMATSU, Jugo
Powerful new weapons are mass-produced to China's own designs (1972) J 358

WALLACE, Henry A.
The US, the UN and Far Eastern agriculture (1950) 830

WANG, Chi
Nuclear research in Mainland China (1967) 359

WANG, K.P.
China (1970/71) 623

WANG, K.P.
Mineral industry of Mainland China (1965, 1967-69) 624

WANG, K.P.
The mineral resource base of Communist China (1968) 654

WANG, K.P.
Mining and metallurgy (1961) 625

WANG, K.P.
Rich mineral resources spur Communist China's bid for industrial power (1960) 626

WANG, Kung-lee
China's mineral industries in 1967: victims of the Cultural Revolution (1969) 627

WANG, K.L.
Mainland China (1968) 628

WANG, Ling and Joseph NEEDHAM 926

WANG, Ling and Joseph NEEDHAM 927

WANG, Ling and Joseph NEEDHAM 928

WANG, Ling, Joseph NEEDHAM and Derek J. PRICE 929

WANG, Ling, Joseph NEEDHAM and Derek J. PRICE 930

WARE, J.R.
Alchemy, medicine and religion in the China of A.D. 320; the Nei P'ien of Ko Hung (1966) 941

WARNER, Charles
Developing science (1964) 396

WAY, E. Leong
Pharmacology (1961) 274

WEI, Anton W.T.
Minerals in China in 1961 (1962) 629

WEI, A.W.T. and Edgar J. GEALY 615

WESTGATE, R.
Industrialization of China moves ahead (1958) 596

WESTGATE, R.W.
Industrialization behind the bamboo curtain (1955) 597

WESTGATE, Robert
Red China claims large oil resources are being developed (1960) 562

WHEELWRIGHT, E.L. and Bruce MCFARLANE
The Chinese road to Socialism (1970) 20

WIENS, Harold J.
Development of geographical science, 1949-1960 (1961) 295

WILGRESS, D.
China's forward leap in science (1960) 206

WILSON, Dick
China's economic prospects (1966) 21

WILSON, D.
China's economic situation (1967) 55

WILSON, D.
China's industrial prospects (1964) 56

WILSON, D.
China's nuclear effort (1965) 361

WILSON, D.
China's trading prospects (1964) 668

WILSON, D.
Peking's trading plans (1965) 669

WILSON, D.
A quarter of mankind (1969) 207

WILSON, D.
Showing off in China (1964) 670

WILSON, D.
Technology in China (1965) 256

WINCHESTER, John
Importance of Chinese for scientific communication (1960) 397

WINFIELD, Gerald F.
China: the land and the people (1948) 831

WOLFLE, Dael
Chinese embargo (1961) 398

WOLFSTONE, Daniel
Burma's honeymoon with China (1961) 765

WOLFSTONE, D.
China's research reactor (1959) 362

WONG, George H.C.
China's opposition to Western science during Late Ming and Early Ch'ing (1963) 832

WONG, Ming and Pierre HUARD 857

WONG, M. and P. HUARD 858

WONG, M. and P. HUARD 859

WONG, M. and P. HUARD 860

WONG, M. and P. HUARD 714

WONG, M. and P. HUARD 861

WOOD, Frank Bradshaw
Astronomy (1961) 270

WORLD OIL
International outlook issue: China (annual) 564

WORLD OIL
Red China's 'lost' oil field is found (1968) 565

WORLD OIL
What's happening around the world: China (1956) 655

WORLD PETROLEUM
Mainland China (1971) 567

WORLD PETROLEUM REPORT
China (annual) 568

WORTH, Robert M.
Health trends in China since the Great Leap Forward (1965) 732

WORTH, R.M.
Strategy of change in the People's Republic of China--the rural health center (1967) 733

WU, Grace Hsia, Y.L. WU and H.C. LING 647

WU, Leng; in Union Research Institute 201

WU, T.Y.
Nuclear physics (1961) 268

WU, Y.-L., H.C. LING and Grace Hsia WU
The spatial economy of Communist China. A study on industrial location and transportation (1967) 647

WU, Yuan-li
China (1969) 630

WU, Y.L.
Chinese industrialization at the crossroads (1961) 22

WU, Y.L.
The economic potential of Communist China (1963) 113

WU, Y.L.
The economy after twenty years (1970) 23

WU, Y.L.
The economy of Communist China; an introduction (1965) 24

WU, Y.L.
Expansion of the Chinese research and development industry (1965) 209

WU, Y.L.
Industrialization under Chinese Communism (1960) 25

WU, Y.L.
The steel industry in Communist China (1965) 612

WU, Y.L.
Future economic development in China (1970) G 26

WU, Y.L. with H.C. LING
Economic development and the use of energy resources in Communist China (1963) 656

WU, Y.L. and Robert SHEEKS
The organization and support of scientific research and development in Mainland China (1970) 210

YABUUCHI, Kyoshi
China's science and Japan (1970) J 942

YABUUCHI, Kiyoshi (Yabuuti, Kiyosi)
Sciences in China from the fourth to the end of the twelfth century (1958) 943

YAMADA, Keiji
Labor, technology and human beings (1972) J 114

YAMADA, K.
Question for the future: the Chinese experiment (1968) J 211

YAMADA, K.
Cultural Revolution and Chinese tradition (1971) F 257

YAMADA, K. and Nakaoka TETSURO
Science suggests China's direction (1972) J 453

YAMADA, K. and J. NEEDHAM 240

YAMAGUCHI, John
Electronics in Communist China (1960) 518

YAMAGUCHI, Tomio
Medical education in the People's Republic of China (1964) J 454

YAMAMOTO, Hideo
Agricultural Revolution and farm production methods in China (1964) J 494

YAMAMOTO, H.
The process of change in the agricultural technology system of China (1965) J 495

YAMANOUCHI, Kazuo
Communist China's dependence upon the Soviet Union in her economic development (1960) J 57

YAMANOUCHI, K., Masahisa SUGANUMA and Reiitsu KOJIMA 109

YAMATO, Shuzo
Present and future of the machine tool industry (1972) J 594

YANG, L.S., John K. FAIRBANK and Alexander ECKSTEIN 783

YAP, Pow-meng
The place of science in China (1944?) 833

YEH, K.C.
Communist China's petroleum situation (1962) 569

YEH, K.C.
Electric power development in Mainland China: prewar and postwar (1956) 503

YEH, K.C.
Soviet and Communist Chinese industrialization strategies (1965) 42

YONG, Lam-lay
The geometrical basis of the Ancient Chinese square-root method (1970) 944

YOUNG, G.B.W.
Mainland China's chemical industry (1965) 531

YOUNG, G.B.W.
Red China (1965) 532

YOUNG, G.B.W.
Some remarks on scientific achievement in Communist China (1962) 213

YU, Arthur
Chemistry (1961) 275

ZAUBERMAN, Alfred
Soviet and Chinese strategy for economic growth (1962) 43

AKABANE, Nobuhisa
Consideration of methodology in the analysis of technological levels (1967) J 58

ASAHI NEWSPAPERS (ed.)
Science and technology in Modern China (1972) J 2

BAUM, Richard
Chinese science (1971) 209a

BRIGHTMAN, R.
Science in China (1949) 774

BROADBENT, K.P. (ed.)
The development of Chinese agriculture 1949-1970 (n.d.) 463

BUCK, John Lossing, Owen L. DAWSON and Yuan-li WU
Food and agriculture in Communist China (1966) 3

CHEMISTRY
Molecular biology in China (1968) 276

CHESNEAUX, Jean
The Chinese miracle (1972) F 836

CLAUSER, H.R.
China's research becoming visible again (1971) 231

DAWSON, Owen L., John Lossing BUCK and Yuan-li WU 3

DEAN, Genevieve C.
China's technological development (1972) 5

DEAN, G.C.
Science, technology and development: China as a 'case study' (1972) 6

DEDIJER, S.
The sixth column (1966) 416

DIRECTORATE FOR SCIENTIFIC AFFAIRS
Chinese scientific and technical literature (1966) 370

DIRECTORATE FOR SCIENTIFIC AFFAIRS
Study on the utilization of scientific and technical literature from China (1965) 371

ECKSTEIN, Alexander, Walter GALENSON and Ta-chung LIU (eds.)
Economic trends in Communist China (1968) 9

ENNIS, Thomas E.
The role of Chinese science and technology in modern civilization (1966) 843

FEINBERG, Betty
Science in China: report on the AAAS symposium (1961) 140

FRASER, Stewart E. (ed.)
Education and Communism in China (1971) 421

GALENSON, Walter, Alexander ECKSTEIN and Ta-chung LIU 9

GOULD, Sidney H.
Sciences in Communist China (1961) 143

ISBERG, Pelle
Natural sciences in China. Sw 149

ISHIKAWA, Shigeru
Choice of technique during the process of Socialist industrialization--comments on Dobbs' 'Chinese Method' (1961) J 27

ISHIKAWA, S.
Long-term prospects for the Chinese economy (1964-1970) J 11

ISIS
Critical bibliography of the history of science and its cultural influences (annual) 862

KARLSSON, Rolf
China's energy resources: a literature survey (1971) Sw 650

KUO, Leslie T.C. and Peter B. SCHROEDER
Communist Chinese monographs in the USDA library (1961) 374

KUO, L.T.C. and P.B. SCHROEDER
Communist Chinese periodicals in the agricultural sciences (1960) 375

LEAR, John
How Red China is taking science to its peasants (1965) 457

LEE, Amy C. and D.C. Dju CHANG
A bibliography of translations from Mainland Chinese periodicals in chemistry, general science and technology, published by US Joint Publications Research Service, 1957-1966 (1968) 378

LIU, Ta-chung, Alexander ECKSTEIN and Walter GALENSON 9

LOW, Ian
Where Chinese science is going (1968) 174a

LUST, John
Index Sinicus: a catalogue of articles relating to China in periodicals and other collective publications, 1920-1955 (1964) 798

MACIOTI, Manfredo
China uses science policy 'to walk on two legs' (1971) 158

MASSACHUSETTS INSTITUTE OF TECHNOLOGY
Current holdings of Communist Chinese journals in the MIT libraries (1960) 380

MASSACHUSETTS INSTITUTE OF TECHNOLOGY LIBRARIES
KWIC index to the science abstracts of China (1960) 381

MIKI, Ken'ichiro
Communist China's petroleum situation (1964) J 545

NAKAYAMA, Shigeru
Kyoto group of the history of Chinese science (1970) 875

NATURE
Science after the Cultural Revolution (1968) 174b

NATURE
Science and society in Ancient China (1948) 877

NATIONAL LENDING LIBRARY FOR SCIENCE AND TECHNOLOGY
List of scientific and technical periodicals received from China (1964) 382

NATIONAL LIBRARY OF MEDICINE
Current holdings of Mainland Chinese journals (1965) 383

NEEDHAM, Joseph
Clerks and craftsmen in China and the West (1970) 889

NEEDHAM, J.
The Grand Titration: science and society in East and West (1969) 892

NEW SCIENTIST
China's progress in computers (1969) 511

NEW SCIENTIST
Chinese colleges to re-open? (1968) 434

NEW SCIENTIST
A look at China's economic potential (1968) 169

NEW SCIENTIST
Mao's ideology in the factories (1969) 96

NEW SCIENTIST
Turning the heat on Chinese scientists (1968) 241

NEWSWEEK
Whither Red China's 'march on science'? (1961) 168

NIELSON, Robert B.
Scientific, academic, and technical research organizations of Mainland China: a selective listing (1965) 170

THE OIL AND GAS JOURNAL
Red China battles for oil status (1967) 551

THE OIL AND GAS JOURNAL
Red China near self-sufficient in oil (1970) 552

THE OIL AND GAS JOURNAL
Turmoil cripples Red Chinese oil (1968) 556

OKA, Takashi
Science in China: the other face (1964) 373a

OLDHAM, C.H.G.
AAAS symposium on Chinese science (1961) 173

OLDHAM, C.H.G.
The Scientific Revolution and China (1964) 183

PIOTROW, Phyllis T. (ed.)
Population and family planning in the People's Republic of China (1971) 748

PRODUCT ENGINEERING
Red China industry nears technical independence from USSR (1961) 588

RESEARCH POLICY LIBRARY, Lund and Research Survey and Planning Organization, New Delhi
Research potential and science policy of the People's Republic of China; a bibliography (1966) 186

RESEARCH SURVEY AND PLANNING ORGANIZATION 186

RIFKIN, Susan B.
On 'contradictions' among academics (a commentary on a workshop) (1972) 15

RIFKIN, S.B.
Science and technology in China's development: report on a workshop held with the aid of the Joint Committee on Contemporary China (1972) 16

SCHRIMPF, Robert
Summary bibliography of works published in China during the period 1950-1960 on the history of the development of Chinese science and technology (1963) F 932

SCHROEDER, Peter B. and Leslie T.C. KUO 374

SCHROEDER, P.B. and L.T.C. KUO 375

SCIENCE MAINICHI
China: indigenous thought and reality (1971) J 104

SCIENCE NEWS LETTER
China's science behind (1961) 445

SCIENCE NEWS LETTER
China's science growing (1966) 446

SCIENCE NEWS LETTER
Minerals found in China (1961) 653

SCIENCE NEWS LETTER
Red China's Michurinism (1954) 249

SCIENCE NEWS LETTER
Scientific Revolution brews in Red China (1965) 181a

SCIENTIFIC AMERICAN
Science in Communist China (1961) 187

SHAPLEY, Deborah
Chinese science, what the China watchers watch (1971) 251

SHIH, Bernadette P.N. and Richard L. SNYDER
International Union List of Communist Chinese serials; scientific, technical and medical, with selected social science titles (1963) 391

SIVIN, Nathan
On China's opposition to Western science during Late Ming and Early Ch'ing (1965) 825

SIVIN, N.
Science and civilization in China (1968) 928a

SIVIN, N.
Science and civilization in China (1972) 928b

SULLIVAN, Walter
A 'Science Revolution' too, under way in China (1968) 174c

SURVEYS AND RESEARCH CORPORATION
Directory of selected scientific institutions in Mainland China (1970) 196

TRETIAK, Daniel
Russian expert in China (1965) 373b

TSAO, Chia Kuei
Bibliography of mathematics published in Communist China during the period 1949-1960 (1961) 392

DIE UMSCHAU IN WISSENSCHAFT UND TECHNIK
Scientific life in the People's Republic of China (1964) G 200

UNION RESEARCH INSTITUTE
Communist China problem research series (1961-1965) 201

US CONGRESS: Joint Economic Committee
An economic profile of Mainland China (1968) 18

US CONGRESS: Joint Economic Committee
People's Republic of China: an economic assessment (1972) 19

US DEPT. OF COMMERCE, Office of Technical Services
Chinese Mainland science and technology (n.d.) 393

US DEPT. OF STATE, Bureau of Intelligence and Research
Nuclear research and technology in Communist China (1964) 357

US LIBRARY OF CONGRESS
Chinese scientific and technical serial publications in the collections of the Library of Congress (1955+1961) 394

VAN DEN BERG-VAN DE GEER, C.A.
China: guide of books, periodicals and records in the library of the Delft Technological University (1965) 395

WALSH, John
Manpower: Senate study describes how scientists fit into scheme of things in Red China, Soviet Union (1963) 204

WANG, Chi
Mainland China organizations of higher learning in science and technology and their publications (1961) 452

WANG, Chi
Nuclear science in Mainland China: a selected bibliography (1968) 360

WHITE, Lynn Jr.
More pieces to the Chinese puzzle (1967) 928c

WORLD PETROLEUM
China's crude output rises to meet needs (1970) 566

WU, Yuan-li, John Lossing BUCK and Owen L. DAWSON 3

YAMANAKA, Akio
On Sino-Japanese academic (physics) communication (1972) J 399

YAO, York Bing
Bibliography of the study on scientific training in institutions of higher learning in Communist China 1958-1964 (n.d.) 455

7865093
3 1378 00786 5093